AF776150

Autonomic Nervous System - Special Interest Topics

Edited by Theodoros Aslanidis and Christos Nouris

Published in London, United Kingdom

Autonomic Nervous System - Special Interest Topics
http://dx.doi.org/10.5772/intechopen.95141
Edited by Theodoros Aslanidis and Christos Nouris

Contributors
Robert L. Drury, Pavol Svorc, Pavol Svorc Jr, Elvan Wiyarta, Nayla Karima, Vidya Chandrasekaran, Mayowa Adeniyi, Moacir Fernandes de Godoy, Michele Lima Gregório, Manuel Víctor López-González, Marc Stefan Dawid-Milner, Amelia Díaz-Casares, Laura Carrillo-Franco, Marta González-García, Tetsuya Tanioka, Miki Sato, Kyoko Osaka, Feni Betriana, Ryuichi Tanioka, Savina Schoenhofer, Irina Alexandrovna Savvina, Anna Olegovna Petrova, Yulia Mikhailovna Zabrodskaya, Theodoros Aslanidis, Christos Nouris

© The Editor(s) and the Author(s) 2022
The rights of the editor(s) and the author(s) have been asserted in accordance with the Copyright, Designs and Patents Act 1988. All rights to the book as a whole are reserved by INTECHOPEN LIMITED. The book as a whole (compilation) cannot be reproduced, distributed or used for commercial or non-commercial purposes without INTECHOPEN LIMITED's written permission. Enquiries concerning the use of the book should be directed to INTECHOPEN LIMITED rights and permissions department (permissions@intechopen.com).
Violations are liable to prosecution under the governing Copyright Law.

cc BY

Individual chapters of this publication are distributed under the terms of the Creative Commons Attribution 3.0 Unported License which permits commercial use, distribution and reproduction of the individual chapters, provided the original author(s) and source publication are appropriately acknowledged. If so indicated, certain images may not be included under the Creative Commons license. In such cases users will need to obtain permission from the license holder to reproduce the material. More details and guidelines concerning content reuse and adaptation can be found at http://www.intechopen.com/copyright-policy.html.

Notice
Statements and opinions expressed in the chapters are these of the individual contributors and not necessarily those of the editors or publisher. No responsibility is accepted for the accuracy of information contained in the published chapters. The publisher assumes no responsibility for any damage or injury to persons or property arising out of the use of any materials, instructions, methods or ideas contained in the book.

First published in London, United Kingdom, 2022 by IntechOpen
IntechOpen is the global imprint of INTECHOPEN LIMITED, registered in England and Wales, registration number: 11086078, 5 Princes Gate Court, London, SW7 2QJ, United Kingdom

British Library Cataloguing-in-Publication Data
A catalogue record for this book is available from the British Library

Additional hard and PDF copies can be obtained from orders@intechopen.com

Autonomic Nervous System - Special Interest Topics
Edited by Theodoros Aslanidis and Christos Nouris
p. cm.

This title is part of the Physiology Book Series, Volume 14
Topic: Human Physiology
Series Editor: Tomasz Brzozowski
Topic Editor: Kunihiro Sakuma
Associate Topic Editor: Kotomi Sakai

Print ISBN 978-1-80355-192-0
Online ISBN 978-1-80355-193-7
eBook (PDF) ISBN 978-1-80355-194-4
ISSN 2631-8261

We are IntechOpen, the world's leading publisher of Open Access books

Built by scientists, for scientists

5,900+
Open access books available

144,000+
International authors and editors

180M+
Downloads

156
Countries delivered to

Our authors are among the
Top 1%
most cited scientists

12.2%
Contributors from top 500 universities

WEB OF SCIENCE™

Selection of our books indexed in the Book Citation Index
in Web of Science™ Core Collection (BKCI)

Interested in publishing with us?
Contact book.department@intechopen.com

Numbers displayed above are based on latest data collected.
For more information visit www.intechopen.com

IntechOpen Book Series
Physiology
Volume 14

Aims and Scope of the Series

Modern physiology requires a comprehensive understanding of the integration of tissues and organs throughout the mammalian body, including the cooperation between structure and function at the cellular and molecular levels governed by gene and protein expression. While a daunting task, learning is facilitated by identifying common and effective signaling pathways mediated by a variety of factors employed by nature to preserve and sustain homeostatic life. As a leading example, the cellular interaction between intracellular concentration of Ca+2 increases, and changes in plasma membrane potential is integral for coordinating blood flow, governing the exocytosis of neurotransmitters, and modulating gene expression and cell effector secretory functions. Furthermore, in this manner, understanding the systemic interaction between the cardiovascular and nervous systems has become more important than ever as human populations' life prolongation, aging and mechanisms of cellular oxidative signaling are utilised for sustaining life. Altogether, physiological research enables our identification of distinct and precise points of transition from health to the development of multimorbidity throughout the inevitable aging disorders (e.g., diabetes, hypertension, chronic kidney disease, heart failure, peptic ulcer, inflammatory bowel disease, age-related macular degeneration, cancer). With consideration of all organ systems (e.g., brain, heart, lung, gut, skeletal and smooth muscle, liver, pancreas, kidney, eye) and the interactions thereof, this Physiology Series will address the goals of resolving (1) Aging physiology and chronic disease progression (2) Examination of key cellular pathways as they relate to calcium, oxidative stress, and electrical signaling, and (3) how changes in plasma membrane produced by lipid peroxidation products can affect aging physiology, covering new research in the area of cell, human, plant and animal physiology.

Meet the Series Editor

Prof. Dr. Thomas Brzozowski works as a professor of Human Physiology and is currently a Chairman at the Department of Physiology and is V-Dean of the Medical Faculty at Jagiellonian University Medical College, Cracow, Poland. His primary area of interest is physiology and pathophysiology of the gastrointestinal (GI) tract, with a major focus on the mechanism of GI mucosal defense, protection, and ulcer healing. He was a postdoctoral NIH fellow at the University of California and the Gastroenterology VA Medical Center, Irvine, Long Beach, CA, USA, and at the Gastroenterology Clinics Erlangen-Nuremberg and Munster in Germany. He has published 290 original articles in some of the most prestigious scientific journals and seven book chapters on the pathophysiology of the GI tract, gastroprotection, ulcer healing, drug therapy of peptic ulcers, hormonal regulation of the gut, and inflammatory bowel disease.

Meet the Volume Editor

Dr. Theodoros K. Aslanidis received an MD from Plovdiv Medical University, Bulgaria, and a Ph.D. from Aristotle University of Thessaloniki, Greece. After serving as a medical doctor in the Hellenic Army Force and as a rural physician at Outhealth Centre, Iraklia and Serres' General Hospital, Greece, he completed anesthesiology specialty training at Hippokratio General Hospital of Thessaloniki. He also completed Critical Care subspecialty training at AHEPA University Hospital, and the Prehospital Emergency Medicine postgraduate program, Hellenic National Centre for Emergency Care. He served as an EMS physician and emergency communication center medic before moving to his current post as consultant-researcher at the Intensive Care Unit, St. Paul General Hospital of Thessaloniki, Greece. He also serves as a senior lecturer in the Research Faculty, College of Offshore and Remote Medicine, Pretty Bay, Malta.

Dr. Christos Nouris received an MD from the School of Medicine, Aristotle University of Thessaloniki, Greece, and an MSc in Medical Research Methodology from the same university. After completing his residency in anaesthesiology at AHEPA University Hospital, he worked as a consultant anaesthesiologist in the District General Hospital of Veria, Greece. Later, he completed his fellowship in intensive care at "G. Papageorgiou" General Hospital, Thessaloniki, Greece. Since 2017 he has been working as a consultant at AHEPA University Hospital. He also teaches medical students at the School of Medicine, Aristotle University of Thessaloniki, and students in the Postgraduate Nursing Specialties Program, University General Hospital AHEPA, and the Committee for the European Education in Anesthesiology (CEEA) teaching programs.

Contents

Preface

The autonomic nervous system (ANS) is a very complex, multilayer neural network that maintains physiologic homeostasis in response to internal and external stimuli. It has both central and peripheral anatomical and physiological integration. ANS can be considered as a subsystem of the nervous system that possesses several levels (groups) for receiving, transmitting, generating, and processing information; an "artificial intelligence" biological system.

This book enriches our current understanding about various ANS subjects. It begins with a short introductory chapter by the editors that reviews the application of electroceuticals therapies in various ANS states.

The first section is the Introduction

The second section includes three chapters. The first chapter, written by Dr. Sato et al., reviews the importance of the ANS in a healthy lifestyle and sleep status for healthy aging. The second chapter, written by Dr. Moacir et al., presents a narrative review of heart rate variability as a homeostasis level index. Finally, the chapter by Dr. Mayowa et al. reviews the literature about the ANS–stress relationship.

The third section includes six chapters. The first chapter, by Prof. Alexandrova et al., discusses general anesthesia and the ANS. The second chapter, by Dr. Wiyarta et al., is about heart ANS basic science and clinical application. The third chapter by Dr. Robert Drury highlights the new emerging technologies for ANS assessment in ambulances. The fourth chapter by Prof. Lopez presents a detailed description of the central control of the larynx in mammals, and the fifth chapter by Prof. Chandrasekaran focuses on signaling pathways that regulate axonogenesis and dendritogenesis in sympathetic neurons. The final chapter, written by Prof. Svorc, describes ANS changes under general anesthesia in rats.

The diversity of the subjects presented in the book make it a valuable reference for future researchers.

Theodoros Aslanidis
Consultant-Researcher,
ICU, St. Paul General Hospital,
Thessaloniki, Greece

Senior Lecturer,
College of Offshore and Remote Medicine,
Pretty Bay, Malta

Christos Nouris
Consultant,
Anesthesiology Department,
AHEPA University Hospital,
Thessaloniki, Greece

Section 1

Introduction

Chapter 1

Introductory Chapter: Electroceuticals of Autonomic Nervous System

Christos Nouris and Theodoros Aslanidis

1. Introduction

The electrical stimulation of human tissue for therapeutic purposes is not a recent concept. Scientists have long ago dallied with the idea of delivering electrical current with implanted devices on human tissue for the development of highly selective organ-specific treatment, instead of using systemic pharmacotherapy. Back on the 70's Braunwald tested electrical stimulation of the carotid sinus nerve to treat cardiac disease [1]. Nowadays, cardiac pacemaker/defibrillator is widely applied as an electrical device used to deliver a current discharge on myocardial tissue in the attempt of setting the pace of cardiac activity or resetting cardiac electrical stability [2].

The nervous system is very appealing for the application of this notion. As an interconnected network of many billions of cells which communicate by sending electrical impulses, the nervous system regulates a vast landscape of organ function. Neuroscience has made a great deal of effort over the last century to understand how information is presented and processed by neuronal circuits in the brain and spinal cord. Neural interface technologies have been used for stimulation and recording. Electrical current applied to specific parts of the nervous system causes neurons to transmit signals to their targets, whereas in other cases electrical stimulation may disrupt the signal being conveyed by nearby neurons. Neurostimulation technology offers new opportunities. In recent years stimulation of neural structures have been used for treatment of neurological disorders and injury. Paralyzed people have been able to move [3], lost hearing has been restored with cochlear implants [4] and Parkinson's disease's symptoms have been alleviated with deep brain stimulation.

Neurostimulation can be delivered selectively to specific parts of the nervous system by several means. Neural structures can be stimulated with electrical current passed through the skin or with non-implantable methods such as transcranial magnetic stimulation, focused ultrasound and high-frequency electrical fields. However, the mainstay of neurostimulation is implantation of electrodes within the body, a process which requires complex surgery on delicate neural tissues. The Central Nervous System (CNS) areas contain anatomically overlapping cell populations with diverse functions. Thus, lowering an electrode into a deep brain structure has the risk of producing non-selective effects and damaging surrounding areas. Autonomic Nervous System, which is used by the brain to influence other organs in the body, is different, though.

2. Specific applications

The Autonomic Nervous System (ANS) comprises neural cell bodies located in CNS and peripheral ganglia. The fibers of ANS neurons get mixed within cranial and spinal peripheral nerves creating afferent and efferent bundles that supply smooth muscle and glands. This way, complex peripheral neural networks that influence the function of internal organs are in intimate association with their target tissues [5]. These peripheral neural networks are located in the thorax and viscera influencing the function of internal organs. Functions controlled by these neural networks include tracheal and bronchial secretions, cardiac activity, arteriole diameter, pancreatic cell secretions, gastrointestinal motility and secretions, urinary bladder contraction, immune system function and mobilization of energy stores from liver and fat [6–8].

The aforementioned explain why ANS peripheral nerves and ganglia are principal targets for electrical neurostimulation therapy. They are typically located peripherally and so are easily accessed by surgery with lower risk of tissue damage. They lack the anatomical and physiological complexity of CNS regions, thus lowering the possibility of non-selective adverse effects. Finally, they are implicated in a large variety of body functions making it appealing to try to produce therapeutic effects by "fine tuning" their complex interactions.

The control of complex neural circuits with electroceutical therapy for treatment of diseases is based on the novel concept of the neural fulcrum. That is, when bioelectric interventions push the autonomic neural networks in one direction, the endogenous reflex control pushes back. The result is that, although there are minimal changes in basal function, the response of the entire network to stress is restrained. This way the hyperdynamic reflex responses, commonly associated with disease progression, are counteracted [9].

Stimulation of organs' neural circuits requires sophisticated implantable electronic devices. These medical devices are called electroceuticals and they incorporate novel biocompatible materials, miniature electronics and computer software to modulate neuronal signaling. Over the past few years research focused on developing electronic devices for selective organ-specific treatment and fewer side-effects. The goal is to produce electroceuticals with grater selectivity and fewer complications [10]. In addition to functionality, device miniaturization, conformability, biocompatibility, and/or biodegradability are the main engineering targets [11].

Electroceutical targeting of the ANS highlights recent advances in the field of electrical neuromodulation and contributes to the treatment of numerous diseases. Targeting of cardiac disease is one clear example. The T_1–T_2 region of the paravertebral chain has been identified as a critical nexus point for sympathetic control of the heart [12]. Application of an electroceutical device to the T_1–T_2 region blocks the sympathetic outflow to the heart without compromising basal cardiac function. In this case electroceuticals aim to restrain the hyperdynamic sympathetic response/withdrawal of central parasympathetic tone which is implicated in the pathophysiology of heart failure [13] and arrhythmias [14]. Electrical stimulation of nerves innervating the carotid sinus activates local baroreceptors resulting in reduced sympathetic outflow and augmented parasympathetic tone providing effective treatment for patients with resistant hypertension [15]. Spinal cord stimulation is also used for pain management [16] and for promoting neural repair and regeneration after injury or for modulating neural plasticity mechanisms that may assist to recover lost functions [17].

Vagus nerve is an important therapeutic target of electroceuticals because it contains afferent and efferent pathways implicated in the communication between brain and abdominal organs. Gastrointestinal tract, pancreas and hepatic portal vein are innervated by vagal branches and their malfunction is linked to gastrointestinal, metabolic and inflammatory diseases. Abdominal vagus nerve stimulation is supported by clinical trials and has been approved for the treatment of gastrointestinal motility disturbances like gastroparesis [18] and for weight loss in moderately obese patients [19]. Cervical vagus nerve stimulation has been shown to reduce inflammatory cytokines and have anti-inflammatory actions [20] that extend for a long period of time after brief stimulation, providing promising results for clinical remission of Crohn's disease [21] and rheumatoid arthritis [22]. Cervical vagus nerve stimulation devices are also used to control epileptic seizures [23] and treatment-resistant depression [24]. In this case the therapeutic mechanism is unclear, but large-scale changes in CNS activity are indicated by functional imaging studies [25]. However, the efficacy of vagus nerve stimulation in the treatment of epilepsy and depression is low (30%), because reduced stimulatory levels are obligatory in order to avoid adverse effects from non-selective stimulation of the cervical vagus trunk.

Bladder dysfunction affects a large population and control of bladder function is implicated in the field of electroceuticals of ANS. Sacral nerve stimulation is approved for treatment of urinary incontinence [26], while the efficacy of the method in neurogenic bladder dysfunction after spinal cord injury is under exploration [27].

Despite the application of electroceuticals of ANS in modern medicine and their participation in the treatment of several pathological conditions with important clinical effects, this technology has still a long way to go. The studies published include a large variety of organ systems, target nerves and diseases and the parameters of electrical stimulations used are diverse regarding amplitude, pulse duration, duty cycle and frequency. These parameters were often developed in animal studies and applied directly to human studies, and they were mostly empirically determined without in depth understanding of the underlying effects of stimulation.

3. Conclusions: perspectives

Electroceutical devices of ANS are unquestionably part of future medicine. An ongoing challenge is to refine electrode position and clarify stimulation patterns in order to achieve increased nerve region and physiological pathways selectivity. Another goal is to develop closed-loop circuits which use sensors designed to quantify neural activity from axons. This dynamic readout will be used by the device's control center to adjust and "tune" the level of neuromodulation. Wireless power delivery to implanted electrodes using focused beam-forming energy to avoid complex or follow-up surgery for the removal of an implanted battery is also under research.

However, the most important parameter, remains the development of a deeper understanding of the underlying biological effects of electrical neurostimulation. Comprehension of precise molecular and physiological changes of ANS during electrical stimulation is the golden key for more effective future electroceutical devices.

Conflict of interests

The author has no conflict of interest.

Author details

Christos Nouris[1]* and Theodoros Aslanidis[2]

1 Anesthesiology Department, AHEPA University Hospital of Thessaloniki, Thessaloniki, Greece

2 Intensive Care Unit, St. Paul General Hospital of Thessaloniki, Thessaloniki, Greece

*Address all correspondence to: xnouris@yahoo.gr

IntechOpen

© 2022 The Author(s). Licensee IntechOpen. This chapter is distributed under the terms of the Creative Commons Attribution License (http://creativecommons.org/licenses/by/3.0), which permits unrestricted use, distribution, and reproduction in any medium, provided the original work is properly cited.

References

[1] Braunwald E, Vatner SF, Braunwald NS, Sobel BE. Carotid sinus nerve stimulation in the treatment of angina pectoris and supraventricular tachycardia. California Medicine. 1970;**112**:41-50

[2] Padera R, Schoen F. Cardiovascular medical devices. Biomaterials Science (Fourth Edition). 2020. pp. 1010-1015

[3] Jackson A, Zimmermann JB. Neural interfaces for the brain and spinal cord—Restoring motor function. Nature Reviews Neurology. 2012;**8**:690-699

[4] Bensmaia SJ, Miller LE. Restoring sensorimotor function through intracortical interfaces: Progress and looming challenges. Nature Reviews Neuroscience. 2014;**15**:313-325

[5] Ardell JL, Andresen MC, Armour JA, Billman GE, Chen PS, Foreman RD, et al. Translational neurocardiology: Preclinical models and cardioneural integrative aspects. The Journal of Physiology. 2016;**594**:3877-3909

[6] Chavan SS, Pavlov VA, Tracey KJ. Mechanisms and therapeutic relevance of neuro-immune communication. Immunity. 2017;**46**:927-942

[7] Furness JB, Stebbing MJ. The first brain: Species comparisons and evolutionary implications for the enteric and central nervous systems. Neurogastroenterology and Motility. 2018;**30**:e13234

[8] Mazzone SB, Undem BJ. Vagal afferent innervation of the airways in health and disease. Physiological Reviews. 2016;**96**:975-1024

[9] Ardell JL, Nier H, Hammer M, Southerland EM, Ardell CL, Beaumont E, et al. Defining the neural fulcrum for chronic vagus nerve stimulation: Implications for integrated cardiac control. The Journal of Physiology. 2017;**595**:6887-6903

[10] Birmingham K, Gradinaru V, Anikeeva P, Grill WM, Pikov V, McLaughlin B, et al. Bioelectronic medicines: A research roadmap. Nature Reviews Drug Discovery. 2014;**13**:399-400

[11] Long Y, Li J, Yang F, Wang J, Wang X. Wearable and implantable electroceuticals for therapeutic electrostimulations. Advanced Science. 2021;**8**(8):2004023

[12] Buckley U, Yamakawa K, Takamiya T, Andrew Armour J, Shivkumar K, Ardell JL. Targeted stellate decentralization: Implications for sympathetic control of ventricular electrophysiology. Heart Rhythm. 2016;**13**:282-288

[13] DiCarlo LA, Libbus I, Kumar HU, Mittal S, Premchand RK, Amurthur B, et al. Autonomic regulation therapy to enhance myocardial function in heart failure patients: the ANTHEM-HFpEF study. ESC Heart Fail. 2018;**5**(1):95-100

[14] Salavatian S, Beaumont E, Longpré JP, Armour JA, Vinet A, Jacquemet V, et al. Vagal stimulation targets select populations of intrinsic cardiac neurons to control neurally induced atrial fibrillation. American Journal of Physiology Heart and Circulatory Physiology. 2016;**311**:H1311-H1320

[15] de Leeuw PW, Bisognano JD, Bakris GL, Nadim MK, Haller H, Kroon AA. DEBuT-HT and Rheos Trial

Investigators. Sustained reduction of blood pressure with baroreceptor activation therapy: Results of the 6-year open follow-up. Hypertension. 2017;**69**:836-843

[16] Foreman RD, Linderoth B. Neural mechanisms of spinal cord stimulation. International Review of Neurobiology. 2012;**107**:87-119

[17] GuillermoGarcía-Alías JV, IgnacioDelgado-Martínez XN. Electroceutical therapies for injuries of the nervous system. In: Handbook of Innovations in Central Nervous System Regenerative Medicine. Amsterdam, Netherlands: Elsevier; 2020. pp. 511-537

[18] Jones MP, Ebert CC, Murayama K. Enterra for gastroparesis. The American Journal of Gastroenterology. 2003;**98**:2578

[19] Morton JM, Shah SN, Wolfe BM, Apovian CM, Miller CJ, Tweden KS, et al. Effect of vagal nerve blockade on moderate obesity with an obesity-related comorbid condition: The recharge study. Obesity Surgery. 2016;**26**:983-989

[20] Pavlov VA, Tracey KJ. Neural circuitry and immunity. Immunologic Research. 2015;**63**:38-57

[21] Bonaz B, Sinniger V, Hoffmann D, Clarençon D, Mathieu N, Dantzer C, et al. Chronic vagus nerve stimulation in Crohn's disease: A 6-month follow-up pilot study. Neurogastroenterology and Motility. 2016;**28**:948-953

[22] Koopman FA, Chavan SS, Miljko S, Grazio S, Sokolovic S, Schuurman PR, et al. Vagus nerve stimulation inhibits cytokine production and attenuates disease severity in rheumatoid arthritis. Proceedings of the National Academy of Sciences of the United States of America. 2016;**113**:8284-8289

[23] US Food and Drug Administration. VNS Therapy System: FDA, Premarket Approval (PMA)—Epilepsy. 1997. Available from: https://www.accessdata.fda.gov/scripts/cdrh/cfdocs/cfpma/pma.cfm?id=P970003S207 [Accessed: November 16, 2021]

[24] Müller HHO, Moeller S, Lücke C, Lam AP, Braun N, Philipsen A. Vagus nerve stimulation (VNS) and other augmentation strategies for therapyresistant depression (TRD): Review of the evidence and clinical advice for use. Frontiers in Neuroscience. 2018;**12**:239

[25] Chae JH, Nahas Z, Lomarev M, Denslow S, Lorberbaum JP, Bohning DE, et al. A review of functional neuroimaging studies of vagus nerve stimulation (VNS). Journal of Psychiatric Research. 2003;**37**:443-455

[26] Herbison GP, Arnold EP. Sacral neuromodulation with implanted devices for urinary storage and voiding dysfunction in adults. Cochrane Database of Systematic Reviews. 2009;**2**:CD004202

[27] Redshaw JD, Lenherr SM, Elliott SP, Stoffel JT, Rosenbluth JP, Presson AP, et al. Protocol for a randomized clinical trial investigating early sacral nerve stimulation as an adjunct to standard neurogenic bladder management following acute spinal cord injury. BMC Urology. 2018;**18**(1):72

Section 2

Autonomous Nervous System, Homeostasis and Stress

Chapter 2

Healthy Lifestyle, Autonomic Nervous System Activity, and Sleep Status for Healthy Aging

Miki Sato, Feni Betriana, Ryuichi Tanioka, Kyoko Osaka, Tetsuya Tanioka and Savina Schoenhofer

Abstract

With the super-aging society, it is important to pay attention to the quality of life of older people so that they can face healthy aging. Lifestyle, particularly exercise, autonomic nervous system activities, and sleep status are factors that affect the quality of aging. This chapter explores how those three variables are related and what strategies can be employed to maintain and enhance these variables to prepare. (1) The combination of healthy lifestyles, adequate physical activity, healthy dietary patterns, moderate alcohol consumption, and nonsmoking were related to the risk of cardiovascular diseases. (2) For older people, being physically active is important to the improvement of their physical and mental functions and keeping them independent and mobile. The increasing HRV after exercise might be caused by increasing vagal tone and decreasing sympathetic activity. (3) To reach healthy aging, people should maintain the proper function of autonomic balance activities. This is important because slowing down the decline in sympathetic status might delay many geriatric complaints. (4) To achieve healthy aging, maintaining a healthy sleep is essential. Thus, the key to a lifestyle that facilitates healthy aging is a balance of regular physical exercise and adequate sleep, which mediates and is mediated by autonomic nervous system activity.

Keywords: autonomic nervous activity, lifestyle, exercise, sleep status, healthy aging, and older people

1. Introduction

The world is experiencing an increasing population of older people. Older people are defined by the World Health Organization as people aged 65 years and older [1]. By 2030, it is predicted that 1 in 6 people will be aged 60 years and over, and between 2020 and 2050 people aged 80 years and older will triple to 426 million people worldwide [2].

Unfortunately, increasing age often occurs along with increasing health problems among many older people. A total of 80% older people were reported to have at least one chronic condition, while 68% of older people have two or more chronic conditions, including hypertension, high cholesterol, diabetes, heart failure, depression, and dementia [3].

Not only chronic diseases, but sleep problems were also frequently reported as a common problem among older people. A study involving 54,722 respondents in 16 countries found that the prevalence of sleep problems among older people in Europe ranged from 16.6% in Italy and Denmark to 31.2% in Poland [4]. Another study by Kim et al. [5] among 3074 older people in South Korea revealed that 29.2% of the participants experience insomnia, while Bhaskar et al. [6] found that 33% of the participants in their study also suffered from chronic insomnia. Insomnia among older people is significantly related to heart disease, anemia, or depression [5], all of which affect healthy aging.

Healthy aging refers to a process of developing and maintaining the functional abilities that allow older people to be and to do things for their well-being [7]. For older people to perform functional abilities, various factors need to be taken into consideration, such as maintaining good sleep and healthy behaviors, including a healthy diet, physical exercise, and refraining from tobacco use [2].

Another factor that contributes to healthy aging is maintaining the balance of the autonomic nervous system (ANS). ANS is a component of the peripheral nervous system that regulates the involuntary physiological process of our body, involving blood pressure, heart rate, respiration, digestion, and sexual arousal [8]. ANS is commonly assessed by heart rate variability (HRV), a measure of the variation in time between each heartbeat [9]. HRV decreases with aging independent of pathological conditions or medication use, potentially suggesting that cardiac autonomic modulation diminishes due to normative aging. Men and women showed similar rates of HRV decline [10].

Understanding how a healthy lifestyle, particularly exercise, ANS, and sleep status are related might assist healthcare professionals to promote healthy aging and prepare older people toward healthy aging.

2. Framework of healthy aging

Healthy aging is characterized by the maintenance of the functional abilities that enable older people to experience well-being [7]. Healthy aging does not mean that older people are free of disease, but the influence of their diseases is minimal that they can still maintain their well-being [7]. The increasing life expectancy resulted in an increasing number of older populations in the world. With super-aging society, it is important to pay attention to the quality of life of older people. Lifestyle particularly exercise, autonomic nervous system activities, and sleep status are factors that affect the quality of aging. This chapter explores how those three variables are related and what strategies can be employed to maintain and enhance these variables to prepare for and achieve healthy aging. (1) The combination of healthy lifestyles, adequate physical activity, healthy dietary patterns, moderate alcohol consumption, and nonsmoking were related to the risk of cardiovascular diseases. (2) Being physically active is important to improve their physical and mental functions and keeping them independent and mobile. (3) To reach healthy aging, people should maintain the proper function of autonomic balance activities. (4) To achieve healthy aging, maintaining a healthy sleep is essential. A healthy sleep involves several dimensions such as adequate sleep duration, good sleep quality, the absence of sleep problems, and appropriate sleep timing. Thus, the key to a lifestyle

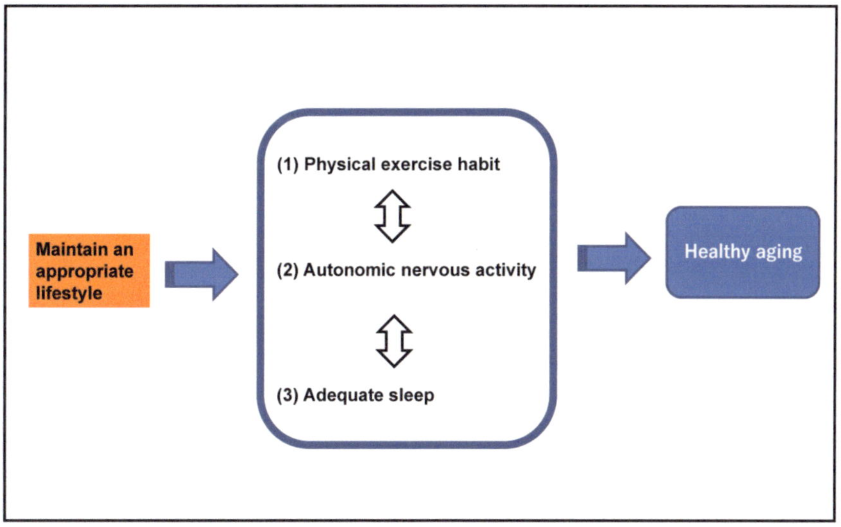

Figure 1.
Relationship of maintaining an appropriate lifestyle for healthy aging as physical exercise habit, autonomic nervous activity, and an adequate sleep pattern [11].

that facilitates healthy aging is a balance of regular physical exercise and adequate sleep, which mediates and is mediated by autonomic nervous system activity.

Healthy aging as a relationship between exercise, ANS activity, and adequate sleep is depicted in **Figure 1**. The key to a lifestyle that facilitates healthy aging is a balance of regular physical exercise and adequate sleep, which mediates and is mediated by ANS activity. In this chapter, the authors explore the relationships among healthy lifestyle, physical exercise habits, ANS activity, and adequate sleep to prepare for healthy aging.

3. Health promotion activities for healthy older people

Table 1 presents the activities that can promote healthy aging and the risk factors that might hinder the achievement of healthy aging.

Maintaining a healthy lifestyle is essential to reaching healthy aging. Some elements of a healthy lifestyle include regular physical exercise [12, 13], maintaining an adequate sleep duration [14], and consuming a healthy diet [15].

No.	Health promotion activities	Risk factors
1.	Physical activity habits	Sedentary lifestyle or having little physical activity
2.	Maintaining an adequate sleep	Sleeping fewer than or more than 7–8 hours
3.	Consuming a healthy diet	Excessive alcohol consumption
4.	Nonsmoking	Smoking cigarettes
5.	Maintaining normal body weight	Obesity or malnutrition

Table 1.
Health promotion activities for healthy older people.

Dietary patterns also play an important role in healthy aging. People with higher diet quality were found to have a higher quality of life score [15]. Healthy dietary patterns such as the Nordic diet and Dietary Approaches to Stop Hypertension (DASH) diet were reported to improve self-rated health and quality of life among older people [15]. The Nordic diet is described as a diet with a higher intake of plant foods, egg, fish, and vegetables [16], and the DASH diet is a plant-focused diet, rich in fruits and vegetables, nuts, with low-fat and non-fat dairy, lean meats, fish and poultry, mostly whole grains, and heart-healthy fats [17]. DASH diet has been proven to decrease cholesterol and blood pressure and is associated with a lower risk of heart disease, diabetes, stroke, kidney stone, and several cancers [17].

The combination of these healthy lifestyles, adequate physical activity, healthy dietary pattern, moderate alcohol consumption, and nonsmoking were related to a 57% lower risk of cardiovascular diseases (CVD) and a 67% lower risk of fatal CVD than performing none or one of these healthy lifestyle behaviors [18].

Conversely, some unhealthy behaviors contribute to poor aging and increased mortality risk, including sedentary lifestyle, excessive alcohol consumption, smoking cigarettes, obesity or malnutrition, sleep duration fewer or more than 7–8 hours, having little physical activity, eating between meals, and not eating breakfast.

4. The importance of exercvise and problems with sitting behavior

Exercise and physical activity are beneficial to people of all ages. Particularly for older people, being physically active is important to the improvement of their physical and mental functions and keeping them independent and mobile [13]. Habitual physical activity might lower cardiovascular mortality among older people [12]. In a longitudinal Cardiovascular Heart Study, which was conducted among 985 older adults for over 5 years, it was found that older people who increased walking pace or walking distance showed favorable HRV indices more than those who decreased walking pace or distance [12]. The finding that exercise improved HRV was also confirmed by another study [19]. Compared with older people who did not perform the exercise, those who performed exercise training three times per week for 16 weeks showed a significantly greater increase in HRV indices [19]. The increasing HRV after exercise might be caused by increasing vagal tone and decreasing sympathetic activity [20].

In relation with physical activity, two conditions that commonly occur to older people are sarcopenia and dynapenia. Sarcopenia is a syndrome characterized by loss of skeletal muscle mass and function, which is related to physical disability, poor quality of life, and death [21]. Alternatively, dynapenia is the age-related loss of muscle strength that is not caused by neurological or muscular disease [22]. A sedentary lifestyle is a major risk factor for chronic disease, frailty, and sarcopenia as well. Physical activity is defined as any movement produced by contracting skeletal muscles that increase energy expenditure [23].

For a person with sarcopenia, it is important to ensure that the patient receives correct and sufficient nutrition and maintains adequate exercise [21]. Resistance training is an effective way to increase muscle mass and strength, regardless of protein supplementation [24].

Although the importance of physical activities has been specified [25], walking activates muscle activity. **Figure 2** shows that the electric potential of each muscle rises during the stance phase. Also, the measured electromyographic (EMG) activities were significantly correlated with the heal-to-toe pressure of the left foot [26]. When

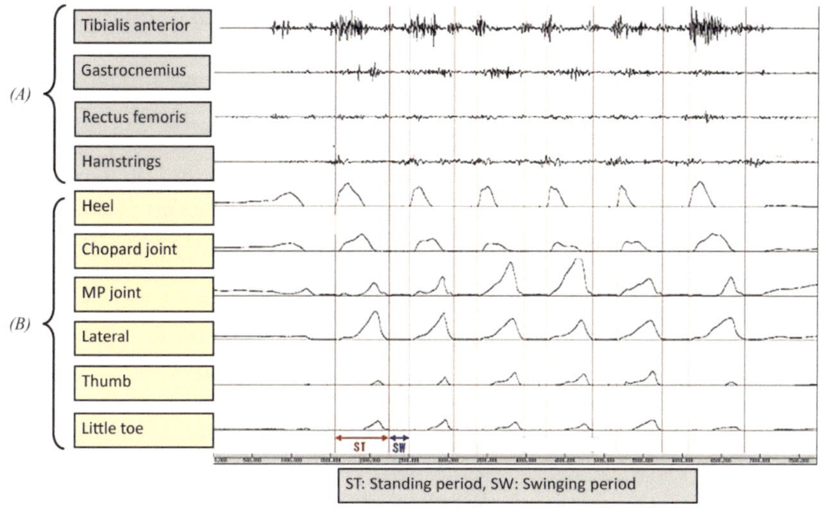

Figure 2.
The typical surface EMG activity recorded during walking freely at 0.5 km/h in a 21-year-old man, the relationship between the sole pressure and lower limb muscle strength. (A) Electromyogram: Tibialis anterior, gastrocnemius, rectus femoris, hamstrings. (B) Sole, pressure of heel-to-toe, pressure: Heel, Chopard joint, metatarsophalangeal (MP) joint, lateral, thumb, little toe. ST: Standing period, SW: Swinging period. This figure was modified based on Dr. Tetsuya Tanioka, his doctoral dissertation in 2001, Kochi University of Technology.

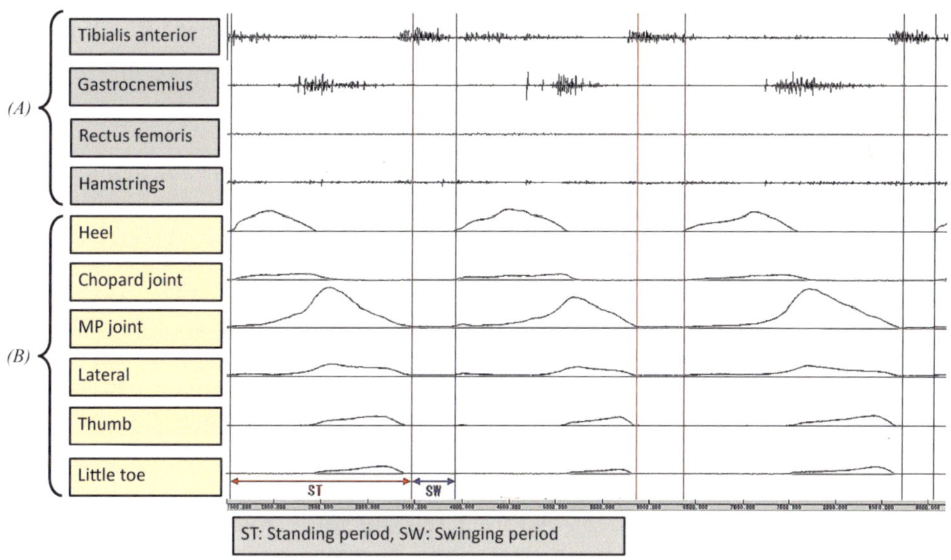

Figure 3.
A 21-year-old man, the relationship between sole pressure and lower limb muscle strength when walking with a walker (rollator) in a 21-year-old man, the relationship between the sole pressure and lower limb muscle strength. (A) Electromyogram: Tibialis anterior, gastrocnemius, rectus femoris, hamstrings. (B) Sole, pressure of heel-to-toe, pressure: Heel, Chopard joint, metatarsophalangeal (MP) joint, lateral, thumb, little toe. ST: Standing period, SW: Swinging period. This figure was modified based on Dr. Tetsuya Tanioka, his doctoral dissertation in 2001, Kochi University of Technology.

people walk, the heel-to-toe pressure shifted serially from the heel to the metacarpophalangeal (MP) joint to toe area during the standing phase, while the pressure was zero during the swinging phase [26]. Physical activity includes daily activities such as standing up from a chair and climbing stairs, as well as walking or biking [23]. As you can see from the figure, these activities are intentional movements for health benefits.

However, it has been reported that older adults exhibit greater interstride dynamic instability of muscle activation patterns during gait [27].

The higher activation for the tibialis anterior, while walking slower might be caused by a deviation from the natural walking pattern of the participant [28]. On the contrary, it has been suggested that walking with a four-wheeled walker (rollator) consistently reduced EMG muscle activity in all lower extremity muscle groups and that increased weight-bearing lead to an increased reduction in muscle activity. Rollator-walking reduces lower-limb muscle activity, but trunk-sway remains unchanged as stability is likely gained through forces generated by the upper limbs [29].

As shown in **Figure 3**. High activation of the tibialis anterior is also observed from the electromyogram when walking with a walker.

From the above information, we can see that walking activates the muscles of the lower limbs.

Lower limb muscle strengthening has an impact beyond reducing the risk of falls because the subjects who performed muscle-strengthening activities showed improvement in other areas such as balance, flexibility, and functional capacity. Lower limb muscle strength training is effective for preventing falls. However, this training should be accompanied by the training of other skills, such as balance and gait, and education [30].

5. Characteristics of ANS activity in older people and healthy aging

The balance of the ANS activity is an important element for healthy aging. In older people, the incidence of many diseases occurs, and chronic diseases often appear along with the dysfunction of the ANS [31].

Sympathetic nervous activities are dominant during the daytime due to human activities, working, tension, and stress. On the other side, parasympathetic nervous activities are dominant during nighttime as people are more relaxed, at rest, and asleep. The sympathetic and parasympathetic nerve is switched as needed to maintain balance in the body [32]. If the ANS is in place, this switching will go smoothly. However, if switching is disturbed, it leads to various physical and mental disorders.

Figure 4 shows that sympathetic activation is associated with daytime work, tension, and stress, and parasympathetic activation is associated with rest, relaxation, and sleep. These autonomic nerve activities must be properly switched and balanced. It is important to activate the sympathetic nerve activity with moderate exercise, leading to the recovery mode of the parasympathetic nerve. It is also important to activate the parasympathetic nerve activity by having relaxing meals and bathing.

In older people, ANS, both sympathetic and parasympathetic activities, commonly change. However, it was reported that varied responses existed among changes in sympathetic parameters with age, including increase, decrease, or unchanged [33]. A cross-sectional study which was conducted among 62 healthy persons in India found that sympathetic and parasympathetic responses declined with the increasing age [33]. Another study revealed that sympathetic nervous activity is significantly found to increase during rest as people age [31].

Furthermore, a study by Fuji et al. [34] found that during awake, sympathetic activity, which is indicated by high-frequency (HF) power/low-frequency (LF) power ratio, showed a positive correlation with human activity, which was measured by actigraph. In contrast, the parasympathetic activity, which was indicated by HF power, increased and showed negative correlations with human activity and sympathetic nervous activity [34]. **Figure 5** shows the typical example of the correlation between sympathetic and parasympathetic activities in a healthy person. Healthy people have high daytime activity values measured by actigraphy and low nighttime

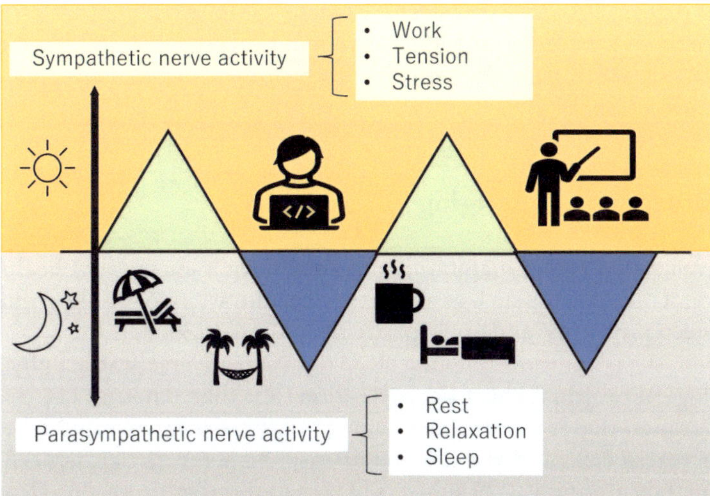

Figure 4.
Sympathetic and parasympathetic activities during daytime and nighttime.

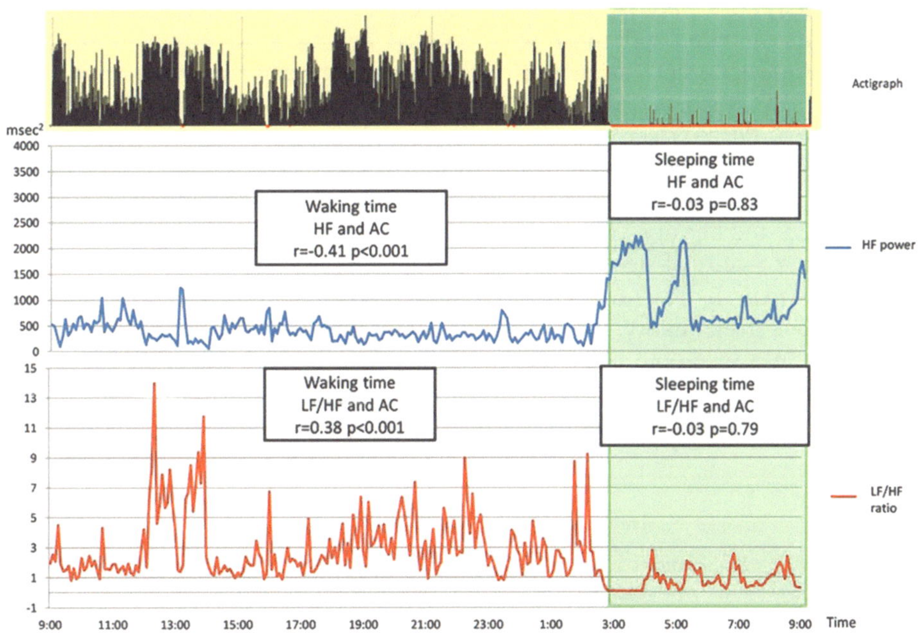

Notes: r, Pearson's r.
Abbreviations: LF, Low frequency power; HF, High frequency power; AC, activity count.

Figure 5.
A typical example of a significant positive correlation between LF/HF and activity count in a healthy person [34].

activity. Sympathetic activity is high during the day in proportion to the activity measured by the actigraph, and parasympathetic activity is high during sleep.

To reach healthy aging, people should keep the balance and maintain the proper function of autonomic balance activities. This is important because slowing down the decline in sympathetic status might delay many geriatric complaints [33].

Two aspects that are frequently mentioned to influence the balance of autonomic activities are physical exercise and sleep. A 1-hour exercise training which is performed three times a week shows improvement in heart rate variability among older people [19]. Meanwhile, poor sleep quality is adversely associated with HRV, heart rate, and blood pressure [35].

6. Sleep status and healthy aging

To achieve healthy aging, maintaining healthy sleep is essential. A healthy sleep involves several dimensions such as adequate sleep duration, good sleep quality, absence of sleep problems, and appropriate sleep timing [36, 37].

Insufficient sleep duration and long sleep duration adversely affect physiological and psychological health. A short sleep duration (less than 6 hours) [14, 38, 39] and long sleep duration that is more than 9 hours [14] are associated with higher prevalence and increased risk of cardiovascular disease, diabetes, stroke, hypertension, and dementia. A study among 10,129 subjects in Iran found that those who slept <6 hours showed a significant risk of CVD, coronary heart diseases (CHD), and hypertension, while those who slept for 8–8.9 hours showed the lowest level of myocardial infarction [38]. Another study among 218,155 participants in Australia found that those who slept <6 hours and > 9 hours had a higher risk of heart diseases, diabetes, stroke, and hypertension compared with those who slept for 7 hours [14].

Not only adequate sleep duration, but the absence of sleep problems is another criterion of healthy sleep. Unfortunately, older people often experience sleep problems due to age-related sleep changes, which result in early waking and fragmented sleep [40]. The common sleep problems, including short sleep duration, poor sleep quality, and later bedtimes are associated with increased food consumption, poor dietary habits, and obesity. Low protein intake was related with difficulty to initiate sleep and poor quality of sleep, while high protein intake and low carbohydrate intake were associated with difficulty to maintain sleep [41].

To improve sleep quality, mood, and quality of life, aerobic physical activity with sleep education about behaviors that help promote sleep can be an effective treatment [42].

Figure 6 shows sleep duration and risk of sleep-related diseases.

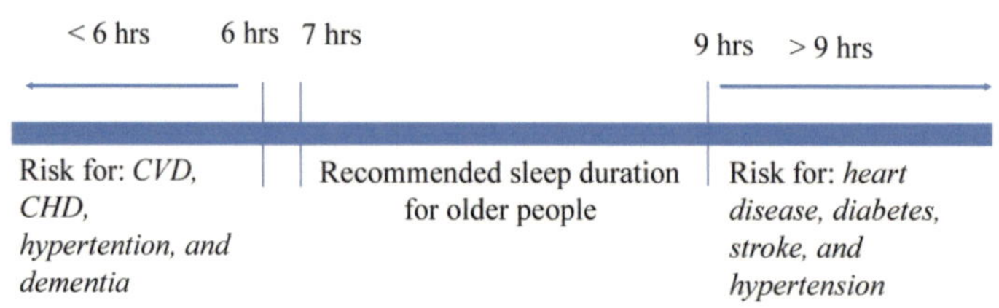

Figure 6.
Sleep duration and risks of sleep-related diseases in older adults.

7. Conclusion

This chapter explored the influence of a healthy lifestyle, physical exercise behavior, ANS activity, and adequate sleep on healthy aging. To reach healthy aging, it is recommended that older people adopt a healthy lifestyle, which involves being physically active, consuming a healthy diet, quitting tobacco, and reducing excessive alcohol use. A healthy lifestyle, particularly physical exercise and healthy sleep improve the balance of ANS activity in older people. The ANS involves important aspects for healthy aging. It is also suggested to maintain healthy sleep with adequate duration of 7–8 hours. Short sleep (<6 hours) and oversleep (>9 hours) increase the risk of CVD, coronary heart disease, diabetes, stroke, and hypertension. With these aspects to keep in mind, healthcare professionals are encouraged to promote activities and a healthy lifestyle to help older people reach healthy aging with good quality of life.

Author details

Miki Sato[1*], Feni Betriana[2], Ryuichi Tanioka[2], Kyoko Osaka[1], Tetsuya Tanioka[3,4] and Savina Schoenhofer[4]

1 Department of Clinical Nursing, Kochi Medical School, Kochi University, Kochi, Japan

2 Graduate School of Health Sciences, Tokushima University, Tokushima, Japan

3 Institute of Biomedical Sciences, Graduate School, Tokushima University, Tokushima, Japan

4 Anne Boykin Institute, Florida Atlantic University, Boca Raton, FL, USA

*Address all correspondence to: m-sato@kochi-u.ac.jp

IntechOpen

© 2022 The Author(s). Licensee IntechOpen. This chapter is distributed under the terms of the Creative Commons Attribution License (http://creativecommons.org/licenses/by/3.0), which permits unrestricted use, distribution, and reproduction in any medium, provided the original work is properly cited. (cc) BY

References

[1] World Health Organization. Mean Ageing and Health: Achieving Health Across the Life Span. 2001. Retrieved from: https://apps.who.int/iris/bitstream/handle/10665/66941/WHO_NMH_NPH_01.2.pdf [Accessed: October 5, 2021]

[2] World Health Organization. Ageing and Health. 2018. Retrieved from: https://www.who.int/news-room/fact-sheets/detail/ageing-and-health [Accessed: October 5, 2021]

[3] National Council on Aging. The Top 10 Most Common Chronic Conditions in Older Adults. 2021. Retrieved from: https://www.ncoa.org/article/the-top-10-most-common-chronic-conditions-in-older-adults [Accessed: October 5, 2021]

[4] van de Straat V, Bracke P.: How well does Europe sleep? A cross-national study of sleep problems in European older adults. International Journal of Public Health 2015;60:643-650. DOI: 10.1007/s00038-015-0682-y

[5] Kim W-H, Kim B-S, Kim S-K, et al. Prevalence of insomnia and associated factors in a community sample of elderly individuals in South Korea. International Psychogeriatrics. 2013;**25**(10):1729-1737. DOI: 10.1017/S1041610213000677

[6] Bhaskar S, Hemavathy D, Prasad S. Prevalence of chronic insomnia in adult patients and its correlation with medical comorbidities. J Family Med Prim Care. 2016;**5**(4):780-784. DOI: 10.4103/2249-4863.201153

[7] World Health Organization. Ageing: Healthy Ageing and Functional Ability. 2020. Retrieved from: https://www.who.int/westernpacific/news/q-a-detail/ageing-healthy-ageing-and-functional-ability [Accessed: October 5, 2021]

[8] Waxenbaum JA, Reddy V, Varacallo M. Anatomy, autonomic nervous system. [Updated 2021 Jul 29]. In: StatPearls [Internet]. Treasure Island (FL): StatPearls Publishing; 2021 Available from: https://www.ncbi.nlm.nih.gov/books/NBK539845/ [Accessed: October 6, 2021]

[9] Campos M. Heart Rate Variability: A New Way to Track Well-Being. 2019. Retrieved from: https://www.health.harvard.edu/blog/heart-rate-variability-new-way-track-well-2017112212789 [Accessed: October 6, 2021]

[10] Jandackova VK, Scholes S, Britton A, Steptoe A. Are changes in heart rate variability in middle-aged and older people normative or caused by pathological conditions? Findings from a large population-based longitudinal cohort study. Journal of the American Heart Association. 2016;**5**(2):e002365. DOI: 10.1161/JAHA.115.002365

[11] Sato M, Betriana F, Tanioka T, et al. Balance of autonomic nervous activity, exercise, and sleep status in older adults: A review of the literature. International Journa of Environmental Research and Public Health. 2021;18(24):12896. DOI: 10.3390/ijerph182412896

[12] Soares-Miranda L, Sattelmair J, Chaves P, et al. Physical activity and heart rate variability in older adults: The cardiovascular health study. Circulation. 2014;**129**(21):2100-2110. DOI: 10.1161/CIRCULATIONAHA.113.005361

[13] McPhee JS, French DP, Jackson D, Nazroo J, Pendleton N, Degens H. Physical activity in older

age: Perspectives for healthy ageing and frailty. Biogerontology. 2016;**17**(3):567-580. DOI: 10.1007/s10522-016-9641-0

[14] Magee CA, Kritharides L, Attia J, McElduff P, Banks E.: Short and long sleep duration are associated with prevalent cardiovascular disease in Australian adults. Journal of Sleep Research 2012;21(4):441-447. DOI: 10.1111/j.1365-2869.2011.00993.x

[15] Govindaraju T, Sahle BW, McCaffrey TA, McNeil JJ, Owen AJ. Dietary patterns and quality of life in older adults: A systematic review. Nutrients. 2018;**26,10**(8):971. DOI: 10.3390/nu10080971

[16] Adamsson V, Reumark A, Cederholm T, Vessby B, Risérus U, Johansson G. What is a healthy Nordic diet? Foods and nutrients in the NORDIET study. Food & Nutrition Research. 2012;**56**:18189. DOI: 10.3402/fnr.v56i0.18189

[17] Heller M. The DASH diet and the mediterranean diet. Retrieved from: https://dashdiet.org [Accessed: October 19, 2021]

[18] Hoevenaar-Blom MP, Spijkerman AM, Kromhout D, Verschuren WM. Sufficient sleep duration contributes to lower cardiovascular disease risk in addition to four traditional lifestyle factors: The MORGEN study. European Journal of Preventive Cardiology. 2014;**21**(11):1367-1375. DOI: 10.1177/2047487313493057

[19] Murad K, et al.: Exercise training improves heart rate variability in older patients with heart failure: A randomized, controlled, single-blinded trial. Congestive Heart Failure 2012;18(4):192-197. DOI: 10.1111/j.1751-7133.2011.00282.x

[20] Routledge FS. Improvements in heart rate variability with exercise therapy. Canadian Journal of Cardiology. 2010;**26**(6):303-312. DOI: 10.1016/S0828-282X(10)70395-0

[21] Santilli V, Bernetti A, Mangone M, Paoloni M. Clinical definition of sarcopenia. Clinical Cases in Mineral and Bone Metabolism. 2014;**11**(3):177-180

[22] Clark BC, Manini TM. What is dynapenia? Nutrition. 2012;**28**(5):495-503. DOI: 10.1016/j.nut.2011.12.002

[23] Rom O, Kaisari S, Aizenbud D, Reznick AZ. Lifestyle and sarcopenia-etiology, prevention, and treatment. Rambam Maimonides Med J. 2012;**3**(4): e0024. DOI: 10.5041/RMMJ.10091

[24] Maltais ML, Ladouceur JP, Dionne IJ. The effect of resistance training and different sources of postexercise protein supplementation on muscle mass and physical capacity in sarcopenic elderly men. Journal of Strength and Conditioning Research. 2016;**30**(6):1680-1687. DOI: 10.1519/JSC.0000000000001255

[25] Hamilton MT, Hamilton DG, Zderic TW. Role of low energy expenditure and sitting in obesity, metabolic syndrome, type 2 diabetes, and cardiovascular disease. Diabetes. 2007;**56**(11):2655-2667. DOI: 10.2337/db07-0882

[26] Tanioka T, Kai Y, Matsuda T, Inoue Y, Sugawara K, Takasaka Y, Tsubahara A, Matsushita Y, Nagamine I, Tada T, Hashimoto F.: Real-time measurement of frozen gait in patient with parkinsonism using a sensor-controlled walker. The Journal of Medical Investigation 2004;51:108-116. DOI: 10.2152/jmi.51.108

[27] Kang HG, Dingwell JB. Dynamics and stability of muscle activations

during walking in healthy young and older adults. Journal of Biomechanics. 2009;**42**(14):2231-2237. DOI: 10.1016/j.jbiomech.2009.06.038

[28] Trinler U, Leboeuf F, Hollands K, Jones R, Baker R. Estimation of muscle activation during different walking speeds with two mathematical approaches compared to surface EMG. Gait & Posture. 2018;**64**:266-273. DOI: 10.1016/j.gaitpost.2018.06.115

[29] Suica Z, Romkes J, Tal A, Maguire C. Walking with a four wheeled walker (rollator) significantly reduces EMG lower-limb muscle activity in healthy subjects. Journal of Bodywork and Movement Therapies. 2016;**20**(1):65-73. DOI: 10.1016/j.jbmt.2015.06.002

[30] Ishigaki EY, Ramos LG, Carvalho ES, Lunardi AC. Effectiveness of muscle strengthening and description of protocols for preventing falls in the elderly: A systematic review. Brazilian Journal of Physical Therapy. 2014;**18**(2):111-118. DOI: 10.1590/s1413-35552012005000148

[31] Hotta H, Uchida S.: Aging of the autonomic nervous system and possible improvements in autonomic activity using somatic afferent stimulation. Geriatrics & Gerontology International 2010;10(S1):S127-S136. DOI: 10.1111/j.1447-0594.2010.00592.x

[32] Bruno B, Valérie S, Sonia P. The vagus nerve in the neuro-immune axis: Implications in the pathology of the gastrointestinal tract. Frontiers in Immunology. 2017;**8**:1452. DOI: 10.3389/fimmu.2017.01452

[33] Parashar R, Amir M, Pakhare A, Rathi P, Chaudhary L. Age related changes in autonomic functions. Journal of Clinical and Diagnostic Research. 2016;**10**(3):CC11-CC15. DOI: 10.7860/JCDR/2016/16889.7497

[34] Fuji S, Tanioka T, Yasuhara Y, Sato M, Saito K, Purnell MJ, et al. Characteristic autonomic nervous activity of institutionalized elders with dementia. Open Journal of Psychiatry. 2016;**6**:34-49

[35] Sajjadieh A, Shahsavari A, Safaei A, et al. The association of sleep duration and quality with heart rate variability and blood pressure. Tanaffos. 2020;**19**(2):135-143

[36] Buysse DJ. Sleep health: Can we define it? Does it matter? Sleep. 2014;**37**(1):9-17. DOI: 10.5665/sleep.3298

[37] Gruber R, Carrey N, Weiss SK, et al. Position statement on pediatric sleep for psychiatrists. Journal of Canadian Academy of Child and Adolescent Psychiatry. 2014;**23**(3):174-195

[38] Yazdanpanah MH, Homayounfar R, Khademi A, et al. Short sleep is associated with higher prevalence and increased predicted risk of cardiovascular diseases in an Iranian population: Fasa PERSIAN Cohort Study. Scientific Reports. 2020;**10**:4608. DOI: 10.1038/s41598-020-61506-0

[39] Sabia S, Fayosse A, Dumurgier J, et al. Association of sleep duration in middle and old age with incidence of dementia. Nature Communications. 2021;**12**:2289. DOI: 10.1038/s41467-021-22354-2

[40] Suzuki K, Miyamoto M, Hirata K. Sleep disorders in the elderly: Diagnosis and management. J Gen Fam Med. 2017;**18**(2):61-71. DOI: 10.1002/jgf2.27

[41] Tanaka E, Yatsuya H, Uemura M, et al. Associations of protein, fat, and carbohydrate intakes with insomnia symptoms among middle-aged Japanese workers. Journal of Epidemiology. 2013;**23**(2):132-138. DOI: 10.2188/jea.je20120101

[42] Reid KJ, Baron KG, Lu B, Naylor E, Wolfe L, Zee PC. Aerobic exercise improves self-reported sleep and quality of life in older adults with insomnia. Sleep Medicine. 2010;**11**(9):934-940. DOI: 10.1016/j.sleep.2010.04.014

Chapter 3

Heart Rate Variability as a Marker of Homeostatic Level

Moacir Fernandes de Godoy and Michele Lima Gregório

Abstract

Many variables have been used as homeostatic level markers. Heart Rate Variability (HRV) has been frequently cited as an indicator of homeostatic status. Low levels of HRV are associated with aging, disease, or increased risk of death. We present a study based on more than 10.5 million data collected from the literature, associating the degree of global clinical impairment of individuals, with their respective HRV data, seeking to establish a classification of Homeostatic Levels. Three specific variables were evaluated: heart rate (HR), the root-mean-square of successive differences between adjacent normal RR intervals in a time interval (RMSSD) and the HF band (HF ms^2). It was possible to detect significant differences between the 83,927 data from healthy individuals and the 382,039 data from individuals with significant homeostatic impairment. It was demonstrated that the RMSSD is very sensitive to the worst homeostatic state, presenting a behavior independent of age and that the values found in the general population do not match the values of apparently healthy individuals. An alphanumeric classification of the homeostatic level in a three-level architecture was proposed, with three stages for each level, which may be extremely useful in prognostic assessment and decision-making about individual people.

Keywords: autonomic nervous system, heart rate variability, homeostatic level

1. Introduction

The human organism is a dynamic, deterministic, non-linear system that shows a sensitive dependence on initial conditions. The amount of cells in the human body is extraordinarily large. A study carried out by Eva Bianconi with collaborators from Italy, Greece and Spain, concluded with the number of $3.72 \pm 0.81 \times 10^{13}$, or approximately 37 trillion cells. There is already an estimate of the amount of cells that need to be removed daily in a healthy human adult, seeking to maintain the body's stability. That number reaches the extraordinary value of 150 billion cells a day! If we remember that the total amount of cells is approximately 37.2 trillion, we conclude that, per day, a healthy human individual loses 0.4% of its cell mass [1].

It is inferred, then, that for the maintenance of life through the proper, harmonious and stable functioning of these cells, in addition to the restoration of lost elements, it is mandatory to spend energy. The clinical concept that refers to this condition of maintenance of conditions of stability is Alostasia. Through Alostasia, Homeostasis is maintained.

The name Homeostasis was created by Walter B. Cannon, in 1932. Literally translated, homeostasis means "staying the same", but this is not entirely accurate. In reality, homeostasis is not a static state; rather, it is a dynamic state.

In biology, homeostasis is classical, the state of steady internal, physical, and chemical conditions maintained by living systems. This is the condition of optimal functioning for the organism and includes many variables, such as body temperature and fluid balance, being kept within certain pre-set limits, and which we will call from now on, as the Homeostatic Level.

One of the fundamental elements for the control of the Homeostatic Level is the Autonomic Nervous System (ANS), with its different components, the sympathetic nervous system, the parasympathetic nervous system and the enteric nervous system [2].

The effects of aging on the autonomic nervous system are multiple and vary between and within both sympathetic and parasympathetic portions. Normal human aging is associated with changes in autonomic control of several bodily functions, particularly those served by cardiovascular and thermoregulatory systems [3].

The assessment of the autonomic nervous system has been possible through the quantification of a biological marker called Heart Rate Variability (HRV). The literature is extremely rich in studies on HRV, and its high applicability in terms of diagnosis and prognosis is a consensus.

It is possible to study HRV in different domains, namely time, frequency and non-linear. In these domains, different variables have already been described, each with its greater or lesser sensitivity.

Briefly, however, we can highlight three of them among those with the greatest clinical applicability: heart rate (HR), the root-mean-square of successive differences between adjacent normal RR intervals in a time interval (RMSSD) and the HF band representing the power in the frequency range between 0.15 and 0.4 Hz (HFms2) [4].

2. Heart rate

Resting heart rate has ceased to be just another vital sign and has become a relevant cardiovascular risk marker. It has long been known that life span is inversely related to resting heart rate in most organisms. The classic article by Levine [5], shows the existence of an inverse semilogarithmic relation between heart rate and life expectancy among mammals, suggesting a predetermined number of heart -beats in a lifetime, with a magic average number of $7.3 \pm 5.6 \times 10^8$ heart-beats/lifetime.

Boudoulas KD et al. [6], make an excellent review relating heart rate, life expectancy and the cardiovascular system. They conclude that many factors regulate heart rate, and it may be these factors, rather than the heart rate itself, which determine survival, but heart rate has multiple direct effects on the cardiovascular system, regardless of the regulatory mechanisms. These effects directly affect the cardiovascular system in multiple ways that, in turn, may affect survival.

From a pathophysiological point of view, the main finding is that resting heart rate is associated with shear and endothelial function in humans [7].

The impact of increased resting heart rate on prognosis is validated in the general population in patients with hypertension, coronary artery disease, or heart failure and irrespective of age, cardiovascular risk factors, or comorbidities, although there is still no definitive confirmation of the prognostic effect of heart rate reduction with the use of drugs such as ivabradine, on primary combined events [8].

3. RMSSD

RMSSD is the root-mean-square of successive differences between adjacent normal RR intervals in a time interval, expressed in millisseconds, and is the primary time domain measure used to assess parasympathetic sources of HRV [4].

Several studies have shown a reduction in RMSSD values in the presence of disease or aging, reflecting a reduction in heart rate variability. Maurer CW et al., in 2016 [9], evaluated the behavior of the autonomic nervous system in 35 patients with functional movement disorders (FMD) compared to 38 healthy controls. They found a significant reduction in RMSSD in patients with FMD ($P = 0.02$), as well as an increased mean heart rate ($P = 0.03$), concluding that decreased vagal tone may reflect increased stress vulnerability in patients with FMD.

DeGiorgio, CM et al. [10], studied 19 subjects with intractable partial seizures, at least three per month, in a randomized clinical trial of omega-3 fatty acids in epilepsy. They looked for whether or not there was a correlation between heart rate variability and the estimated risk of Sudden Unexplained Death in Epilepsy, quantified by the SUDEP-7 Inventory. They found that the RMSSD was inversely correlated with the SUDEP-7 score, $r = -0.64$, $p = 0.004$. Subjects with higher SUDEP-7 scores had reduced levels of HRV (RMSSD). Other time-dependent measures of HRV (SDNN, SDANN) were not significantly correlated with SUDEP risk scores.

In another study, Maheshwari A et al. [11] evaluated a large group of 12,543 individuals from the general population, participating in The Atherosclerosis Risk in Communities Study. They were looking for a relationship between low HRV and sudden cardiac death (SCD). During a median follow-up of 13 years, 215 SCDs were identified. In the group in which sudden deaths occurred, there was a statistically significant difference in heart rate (70.3 ± 13.8 bpm versus 67.7 ± 10.3 bpm; $P = 0.008$) and in HF power ms2 (1.6 ± 1.5 Ln versus 2.1 ± 1.3 Ln; $P < 0.0001$). As for the RMSSD, there was no statistically significant difference between the groups, but in both conditions, the values were below the ideal values for normality (27.3 ± 28.3 ms versus 29.2 ± 23.3 ms; $P = 0.25$).

Based on the knowledge that sepsis is associated with marked alterations in hemodynamic responses, autonomic dysfunction and impaired vascular function, Bongiorno Junior et al. [12], explored the prognostic utility of cardiac output (CO), stroke volume (SV), indices of vagal modulation (RMSSD and SD1), total heart rate variability (HRV) and flow-mediated dilation (FMD) of the brachial artery (%FMD) in 60 patients recruited at an intensive care unit. They found that in the group of 39 patients who did not survive, HR was higher (105 ± 27 bpm versus 84 ± 15 bpm; $P = 0.02$) and it was observed that the RMSSD and SD1 indices could be predictors of endothelial function and RMSSD could predict the risk of death in these patients.

The ROC Curve of RMSSD was useful in predicting 28-day mortality in patients with sepsis. The area under the curve was 0.784 (0.656–0.881). The value of 10.8 ms was chosen as the cut-off point for RMSSD (sensitivity of 77.1%, specificity of 73.9%, the positive likelihood ratio of 2.96 and negative likelihood ratio of 0.31. With RMSSD ≤10.8 ms, the mean survival time was 23.1 days and with RMSSD>10.8 ms, the mean survival time was 23.1 days).

4. HF ms^2

There are three main spectral components in an HRV spectrum named as high frequency (HF), low frequency (LF), and very low frequency (VLF) bands. The HF band represents the power in the frequency range between 0.15 and 0.4 Hz. HF power is generally believed to represent respiration-linked changes in heart rate and is generally accepted as a measure of respiratory sinus arrhythmia (RSA), or the parasympathetic contribution to HRV. RSA refers to the acceleration in heart rate that occurs during inspiration (due to the cardiovascular control center's inhibition of vagal outflow) and the subsequent heart rate deceleration that occurs during expiration, due to vagal restoration [13, 14].

Doheny et al. in 2015 [15], evaluated the possibility of using a non-invasive biomarker that allows early detection of patients at risk of necrotizing enterocolitis (NEC), that is an acute neonatal inflammatory disease that may lead to intestinal necrosis, multi-system failure and death. For that, they used the high frequency (HF) component of heart rate variability. They studied 70 stable preterm infants (gestational age 28-35 week). HF ms^2 was 21.5 ± 2.7 ms^2 in infants that remained healthy and 3.9 ± 0.81 ms^2 in those that later developed stage 2 + NEC ($P < 0.001$). The cut-off value in the ROC curve was 4.68ms^2, predictive for developing NEC with sensitivity and specificity of 89% and 87%, and positive and negative predictive values of 50% and 98%, respectively. They concluded that HF ms^2 may serve as a potential, non-invasive predictive biomarker of NEC-risk in infants.

In 2004, Abramkin et al. [16], studied 188 patients to compare the prognostic value of different noninvasive reflex tests on days 4-11 of myocardial infarction. The age varied from 34 to 75 years, 68% were men, and 93.6% were on beta-blockers, all without heart failure NYHA IV on the day of tests. HF power < 65 ms2 during active standing (OR 28.8, 95% CI 4.1-104.2; $p = 0.001$, positive predictive value 29.4%) was an independent predictor of sudden cardiac death.

In a meta-analytic study carried out in 2021 by Heimrich et al. [17], the objective was to verify whether the analysis of heart rate variability could indicate decreased parasympathetic tone in patients with Parkinson's disease. A total of 47 studies were evaluated, including 2772 individuals, 1566 of which had Parkinson's Disease (65.0 ± 0.6 years) and 1206 were healthy controls (62.6 ± 1.0 years). Based on 24 studies, it was possible to detect that the FH ms2 was significantly lower in the group of patients with the disease (145.2 ± 41.1 versus 219.4 ± 48.8 ms2; $P = 0.002$; heterogeneity 91%).

5. Objective

Considering that the heart rate (HR), the root-mean-square of successive differences between adjacent normal RR intervals (RMSSD) and the HF band power in the frequency range between 0.15 and 0.4 Hz (HFms2), can help to differentiate the homeostatic level between individuals with severe impairment and high risk, and healthy individuals, we performed an intensive review of the literature by collecting published data involving the aforementioned variables, in search of a cutoff value for defining homeostatic reference levels and creating an individualized diagnostic coding.

6. Method

Based on research projects linked to FAPESP - Brazil (2017/12529-7) and CNPq -Brazil (308,555/2018-0), studies involving the use of one or more of the three variables mentioned above were evaluated. In total, it was possible to analyze 164 studies involving the heart rate of individuals with importantly compromised homeostatic level (HL_ic), 181 studies involving the heart rate of apparently healthy individuals (HL_ah), 179 studies involving the RMSSD of individuals with HL_ic, 221 studies involving the RMSSD from HL_ah subjects, 125 studies involving the HF ms2 of subjects with HL_ic and 155 studies involving the HF ms2 of HL_ah subjects. Obviously, there were concurrent studies in certain situations. Due to a large number of references, they are cited separately and available in a *supplementary file*.

7. Statistical analysis

Data were presented as mean and standard deviation, weighted mean, quantities, percentages and correlation coefficients. Comparisons between groups were made by analysis of variance or the Kruskal-Wallis test and its post-tests, according to the indication. Correlation graphs were constructed and Box-Whisker graphs were used for illustration. An alpha error of 5% was accepted, with P values less than or equal to 0.05 being considered significant. The statistical software used was StatsDirect version 3.3.5 (03/22/2021).

8. Results

The total amount of data analyzed was extremely high. **Table 1** below indicates the amounts for each variable under conditions of significantly compromised and apparently healthy homeostasis, as well as the mean and standard deviation values for the age of the group, the mean and standard deviation of the variable, and the weighted mean of the variable.

	N References	N Data	Age [Mean ± SD]	Variable [Mean ± SD]	Variable [weighted mean]	*P*-value
HR HL_ic bpm	164	365,195	48.7 ± 20.7	97.6 ± 20.1	85.7	
HR HL_ah bpm	181	22,443	32.9 ± 21.7	75.9 ± 18.9	69.7	$P < 0.0001$
RMSSD HL_ic ms	179	10,014	54.4 ± 16.5	22.8 ± 19.9	27.4	
RMSSD HL_ah ms	221	35,531	30.1 ± 19.3	45.6 ± 18.8	32.5	$P < 0.0001$
HF ms^2 HL_ic ms^2	125	6830	54.2 ± 15.2	173.7 ± 181.4	155.9	
HF ms^2 HL_ah ms^2	155	25,953	32.7 ± 19.7	565.2 ± 459.1	468.1	$P < 0.0001$

Table 1.
Distribution of the number of studies included, amount of data per variable, according to homeostatic level (importantly compromised [HL_ic] or apparently healthy [HL_ah].

As the behavior of heart rate variability is related to age, linear correlation calculations were made between age (predictor) and the variable to be predicted (HR, RMSSD or $HFms^2$) in the HL_ic and HL_ah groups (**Table 2**; **Figures 1–6**).

Variable	Simple linear regression	Correlation coefficient (r)	Correlation coefficient (r^2)	Two sided P-values
HR HL_ic	−0.493334 age + 121.57553	−0.532497	0.283553	< 0.0001
HR HL_ah	−0.444059 Age + 90.500925	−0.510337	0.260444	< 0.0001
RMSSD HL_ic	−0.159605 Age + 31.493563	−0.131765	0.017362	0.0787#
RMSSD HL_ah	−0.520603 Age + 61.235545	−0.534807	0.286018	< 0.0001
HFms2 HL_ic	−2.572467 Age + 313.130404	−0.215726	0.046538	0.0157
HFms2 HL_ah	−11.218318 Age + 932.609833	−0.480536	0.230915	< 0.0001

#Correlation coefficient is not significantly different from zero.

Table 2.
Distribution of simple linear regression, correlation coefficients (r and r^2) and two-sided P-values, by homeostatic condition.

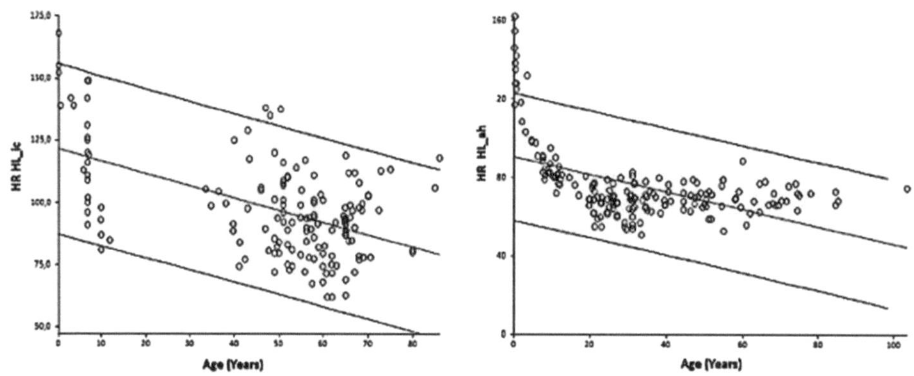

Figure 1.
Correlation graphs (age x heart rate) in the groups of individuals with importantly compromised homeostatic level (HL-ic) and apparently healthy (HL_ah).

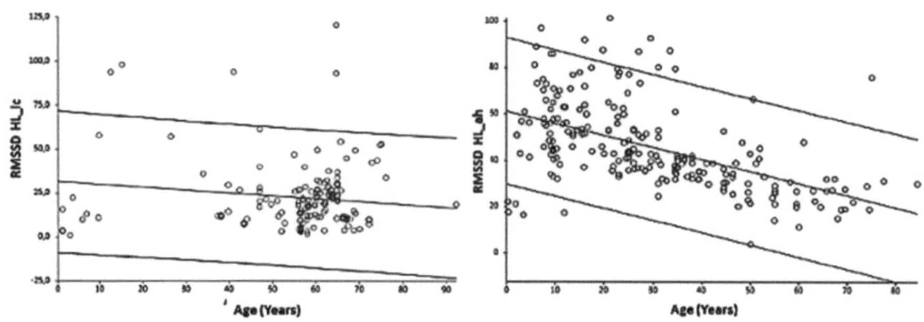

Figure 2.
Correlation graphs (age x RMSSD) in the groups of individuals with importantly compromised homeostatic level (HL-ic) and apparently healthy (HL_ah).

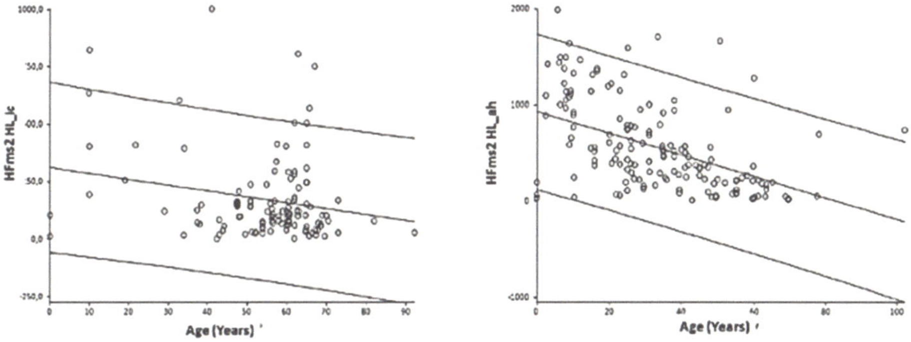

Figure 3.
Correlation graphs (age x HF ms²) in the groups of individuals with importantly compromised homeostatic level (HL-ic) and apparently healthy (HL_ah).

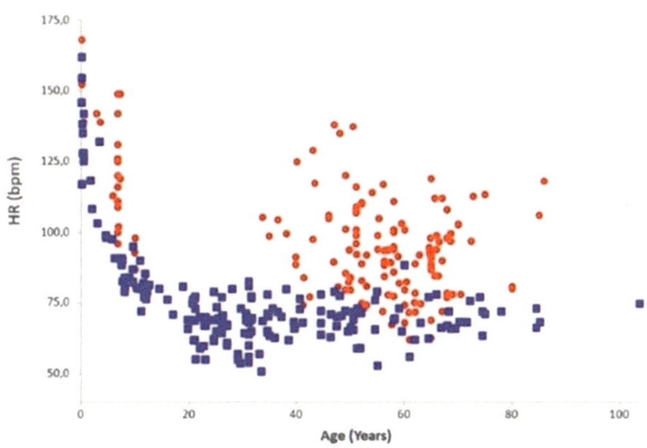

Figure 4.
Scattergram (HR in bpm) for the groups of individuals with importantly compromised homeostatic level (HL- ic; red circles) and apparently healthy (HL_ah; blue squares).

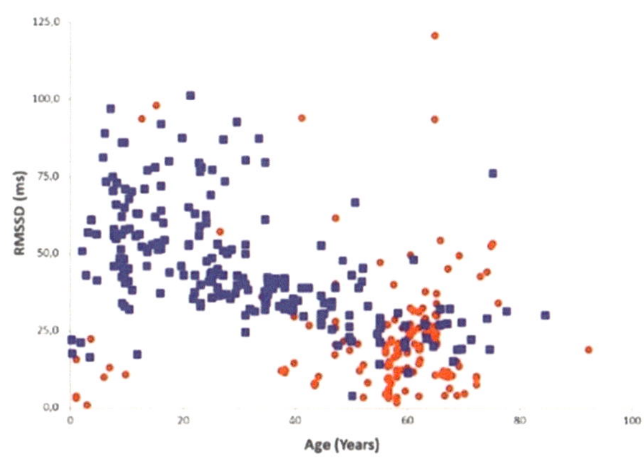

Figure 5.
Scattergram (RMSSD in ms) for the groups of individuals with importantly compromised homeostatic level (HL-ic; red circles) and apparently healthy (HL_ah; blue squares).

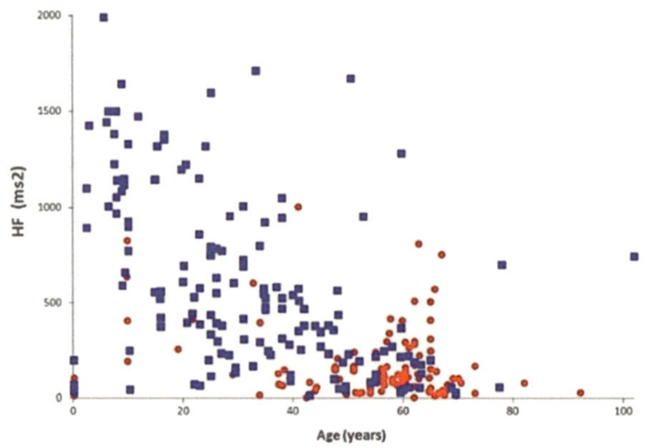

Figure 6.
Scattergram (HF in ms²) for the groups of individuals with importantly compromised homeostatic level (HL- ic; red circles) and apparently healthy (HL_ah; blue squares).

A moderate negative correlation was found between heart rate and age, both in cases with significant homeostatic impairment and in apparently healthy cases. There was also a moderate negative correlation between RMSSD and age, and between HFms^2 and age in the apparently healthy group. The fact that there was only a weak negative correlation between HFms2 and age in the group with significant homeostatic impairment and also the absence of correlation between RMSSD and age in this impaired group, was noteworthy. This may suggest that RMSSD is a more effective or sensitive biological marker of homeostasis, revealing changes regardless of age.

It became also relevant to evaluate the data of the three selected variables, in the group composed of individuals named as being from the general population (HL_gp). Thus, from the global data survey carried out, a number of 10,121,910 were obtained from individuals from the general population, in different age groups. The values of mean, standard deviation, weighted mean, mean age ± standard deviation and number of articles consulted are found in **Table 3** and **Figure 7**.

Comparative statistical analysis between the 3 groups (HL_ah, HL_ic and HL_gp) for the three selected variables, using the Kruskal-Wallis test with post-test Dwass-Steel-Chritchlow-Fligner, showed a non-significant difference between HR HL_ah versus HR HL_gp ($P = 0.6228$); the statistically significant difference between HR HL_ah versus HR HL_ic ($P < 0.0001$); the statistically significant difference between HR HL_gp versus HR HL_ic ($P < 0.0001$).

	Heart rate	RMSSD	HF ms^2
Data (N)	144,817	5,098,117	4,878,976
Mean	71.1	30.3	273.1
Standard Deviation	5.8	12.5	266.2
Weighted mean	68.3	43.2	569.2
Age (mean ± SD)	51.0 ± 15.3	48.9 ± 16.8	50.9 ± 13.8
References	110	138	118

Table 3.
Data and values were obtained in the assessment of the general population (HL_gp).

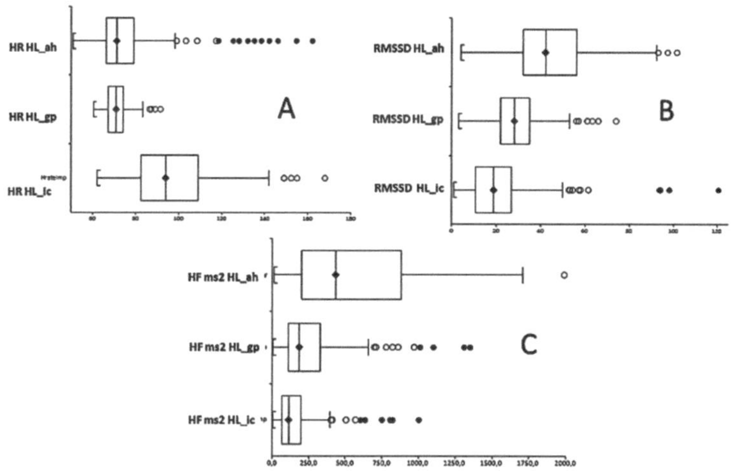

Figure 7.
Box-whisker graphs of the distributions of values for heart rate (A), RMSSD (B) and HF ms2 (C) variables, by the homeostatic level group.

Regarding the variable RMSSD, there was a statistically significant difference between RMSSD HL_ah versus RMSSD HL_gp ($P < 0.0001$); the statistically significant difference between RMSSD HL_ah versus RMSSD HL_ic ($P < 0.0001$); the statistically significant difference between RMSSD HL_gp versus RMSSD HL_ic ($P < 0.0001$).

In the comparative analysis of the variable HF ms2, there was a statistically significant difference between RMSSD HL_ah versus RMSSD HL_gp ($P < 0.0001$); statistically significant difference between RMSSD HL_ah versus RMSSD HL_ic ($P < 0.0001$); statistically significant difference between RMSSD HL_gp versus RMSSD HL_ic ($P = 0.0002$).

Therefore, it is concluded that data from the so-called general population are not suitable to be considered as a normal condition and this must be taken into account when this group is used as a control group.

Finally, based on the weighted average of the results in **Table 1**, on the scatter plots involving the group of individuals with significant homeostatic impairment and the group of apparently healthy individuals, we propose a classification model for the individual homeostatic level. This classificatory model is a three-level, three-stage alphanumeric coding, designed as follows:

Level A: Heart Rate (bpm)
- Stage A1: Heart Rate less than 70 bpm
- Stage A2: Heart Rate between 70 and 85 bpm
- Stage A3: Heart Rate above 85 bpm

Level B: RMSSD (ms)
- Stage B1: RMSSD above 32 milliseconds.
- Stage B2: RMSSD between 32 and 28 milliseconds.
- Stage B3: RMSSD less than 28 milliseconds.

Level C: HF ms^2
- Stage C1: HF ms^2 above 468 ms^2
- Stage C2: HF ms^2 between 468 and 156 ms^2.
- Stage C3: HF ms^2 less than 156 ms^2.

Thus, a totally healthy individual, with an excellent Homeostatic Level and, therefore, with very low risk, would receive the A1B1C1 classification. An individual with a

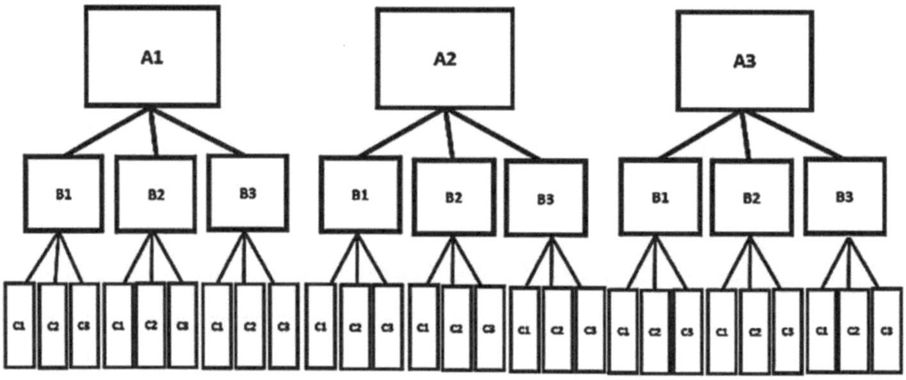

Figure 8.
Set of possibilities in the alphanumeric classification of the individual homeostatic level (Created by the authors).

high basal heart rate, a very low RMSSD value and a very low HF power value would be classified as A3B3C3 indicating high severity, low homeostatic level and, therefore, at high risk. Several intermediate combinations would be possible characterizing the current state of each case. The figure below illustrates the full set of possibilities (**Figure 8**).

In conclusion, the present analytical study, based on an extensive amount of data published in the literature (more than 10.5 million values), referring to three recognized variables of heart rate variability markers of the level of homeostasis, allowed us to define cut-off levels indicative of apparently healthy or with important homeostatic compromise. It was possible to conclude that values obtained in the general population are not equivalent to normal values, a fact that must be considered when this group is used as a control. It was also possible, to elaborate a very simple alphanumeric classification with practical applicability in the characterization of the individual homeostatic level.

Author details

Moacir Fernandes de Godoy[1*] and Michele Lima Gregório[2]

1 Moacir Fernandes de Godoy, Sao Jose do Rio Preto Medical School (FAMERP), São José do Rio Preto, Brazil

2 Michele Lima Gregório, Sao Jose do Rio Preto Medical School (FAMERP), São José do Rio Preto, SP, Brazil

*Address all correspondence to: mf60204@gmail.com

IntechOpen

© 2022 The Author(s). Licensee IntechOpen. This chapter is distributed under the terms of the Creative Commons Attribution License (http://creativecommons.org/licenses/by/3.0), which permits unrestricted use, distribution, and reproduction in any medium, provided the original work is properly cited.

References

[1] Bianconi E, Piovesan A, Facchin F, Beraudi A, Casadei R, Frabetti F, et al. An estimation of the number of cells in the human body. Annals of Human Biology. 2013;**40**(6):463-471. DOI: 10.3109/03014460.2013.807878 Epub 2013 Jul 5. Erratum in: Ann Hum Biol. 2013 Nov-Dec;40(6):471

[2] Deutekom AW. The origins of children's Energy balance-related behavior and physical fitness (thesis). Amsterdam, Netherlands: Vrije Universiteit; 2017

[3] Kuchel GA, Hof PR. Autonomic nervous system in old age. Basel: Karger; 2004

[4] Vanderlei LC, Pastre CM, Hoshi RA, Carvalho TD, Godoy MF. Basic notions of heart rate variability and its clinical applicability. Revista Brasileira de Cirurgia Cardiovascular. 2009;**24**(2):205-217. DOI: 10.1590/s0102-76382009000200018

[5] Levine HJ. Rest Heart Rate and Life Expectancy. Journal of the American College of Cardiology. 1997;**30**(4):1104-1106

[6] Boudoulas KD, Borer JS, Boudoulas H. Heart rate, life expectancy and the cardiovascular system: Therapeutic considerations. Cardiology. 2015;**132**:199-212. DOI: 10.1159/000435947

[7] Fox BM, Brantley L, White C, Seigler N, Harris RA. Association beween resting heart rate, shear and flow-mediated dilation in healthy adults. Experimental Physiology. 2014;**99**:1439-1448. DOI: 10.1113/expphysiol.2014.080960

[8] Custodis F, Reil J-C, Laufs U, Böhm M. Heart rate: A global target for cardiovascular disease and therapy along the cardiovascular disease continuum. Journal of Cardiology. 2013;**62**:183-187. DOI: 10.1016/j.jjcc.2013.02.018

[9] Maurer CW, Liu VD, LaFaver K, Ameli R, Wu T. Impaired resting vagal tone in patients with functional movement disorders. Parkinsonism & Related Disorders. 2016;**30**:18-22. DOI: 10.1016/j.parkreldis. 2016.06.009

[10] DeGiorgio CM, Miller P, Meymandi S, Chin A, Epps J, Gordon S, et al. RMSSD, a measure of vagus-mediated heart rate variability, is associated with risk factors for SUDEP: the SUDEP-7 Inventory. Epilepsy & Behavior. 2010;**19**(1):78-81. DOI: 10.1016/j.yebeh.2010.06.011), 10.1016/j.yebeh.2010.06.011)

[11] Maheshwari A, Norby FL, Soliman EZ, Adabag S, Whitsel EA, Alonso A, et al. Low heart rate variability in a 2-minute electrocardiogram recording is associated with na increased risk of sudden cardiac death in the general population: the atherosclerosis risk in communities study. PLoS One. 2016, 2016;**11**(8):e0161648. DOI: 10.1371/journal.pone.0161648

[12] Bonjorno Junior JC, Caruso FR, Mendes RG, da Silva TR, Biazon TMPC, Rangel F, et al. Noninvasive measurements of hemodynamic, autonomic and endothelial function as predictors of mortality in sepsis: A prospective cohort study. PLoS One. 2019;**14**(3):e0213239. DOI: 10.1371/journal.pone 0213239. Erratum in: PLoS One. 2019 Apr 30;**14**(4):e0216505

[13] Task Force of the European Society of Cardiology and the North American

Society of Pacing and Electrophysiology. Heart rate variability: standards of measurement, physiological interpretation and clinical use. Circulation. 1996;**93**(5):1043-1065

[14] Lee TB, Nicolaas C, Piet B, Margaretha V. Validity of commonly used heart rate variability markers of autonomic nervous system function. Neuropsychobiology. 2019;**8**(1):1-13. DOI: 10.1159/000495519

[15] Doheny KK, Palmer C, Browning KN, Jairath P, Liao D, He F, et al. Diminished vagal tone is a predictive biomarker of necrotizing enterocolitis-risk in preterm infants. Neurogastroenterology and Motility. 2014;**26**:832-840. DOI: 10.1111/nmo.12337

[16] Abramkin DV, Iavelov IS, Gratsianskiĭ NA. Neinvazivnye serdechno-sosudistye reflektornye testy i prognoz vnezapnoĭ serdechnoĭ smerti posle perenesennogo infarkta miokarda: kakoĭ metod predpochest'? [Simple cardiovascular reflex tests in prediction of sudden death after myocardial infarction: Which method to prefer?]. Kardiologiia. 2004;**44**(10):4-12 Russian

[17] Heimrich KG, Lehmann T, Schlattmann P, Prell T. Heart rate variability analyses in parkinson's disease: A systematic review and meta-analysis. Brain Sciences. 2021;**11**:959. DOI: 10.3390/brainsci11080959

Chapter 4

Impacts of Environmental Stressors on Autonomic Nervous System

Mayowa Adeniyi

Abstract

Stress can be described as the perception of discomforts physically, psychologically, or physico-psychologically. During stress, the perceived discomfort indicates there is a deviation from homeostasis. In stress, there is a nonspecific physiological response to stressors, a group of stress-inducing phenomena. Stress-inducing phenomena can be defined as environmental insults, such as perturbed levels of light, temperature, chemicals, ambient oxygen, and noise. Response to stress occurs via the chemical messenger-mediated sympathetic nervous system including the autonomic-adrenal axis. Furthermore, the chemical messenger-mediated sympathetic nervous system determines nonhormonal effects which are often devised as general stress markers. Examples of general stress markers include changes in heart rate, heart rate variability, blood pressure, body temperature, blood glucose, baroreflex sensitivity, among others.

Keywords: stress, autonomic nervous system, hormonal stress markers, nonhormonal stress markers

1. Introduction

The autonomic nervous system is a nervous control saddled with the regulation of vegetative functions, moderating homeostatic processes in the body. It mediates growth, reproduction, digestion, respiration, blood pressure, heart rate, thermoregulation, osmoregulation, penile tumescence, maintenance of glucose, and electrolyte balance in the body. It connects the peripheral organs in the body with the central nervous system. The autonomic nervous consists of the parasympathetic nervous system and the sympathetic nervous system.

The parasympathetic nervous system plays important role in the homeostasis and restoration of homeostasis. It is evoked during rest and relaxation. Parasympathetic signals are transmitted from one neural cell to another cell through cholinergic nerve endings. Examples of such nerve endings include preganglionic cholinergic nerve fibers, postganglionic cholinergic nerve fibers, and preganglionic sympathetic cholinergic nerve fibers and postganglionic sympathetic cholinergic nerve fibers. The acetylcholine released by parasympathetic nerve fibers binds with muscarinic receptors (m_1-m_5) to elicit specific responses.

In some cases, parasympathetic activities are not directly mediated by acetylcholine but by other chemical messengers. For instance, the cholinergic nerve endings to

 IntechOpen

cavernosal arterioles increase the expression of nitric oxide synthase leading to the secretion of nitric oxide causing vasodilation and increased blood flow to penile erectile tissues. Vagal nerve endings in the stomach secrete nitric oxide leading to expansion of the stomach before the arrival of bolus (receptive gastric relaxation) and relaxation of the stomach in the presence of bolus (adaptive gastric relaxation).

The parasympathetic nervous system originates from the cranial nerves and sacral segments of the spinal cord (craniosacral outflow). Usually, the preganglionic parasympathetic cell bodies are situated in the cranial nerves and sacral segments of the spinal cord. However, the parasympathetic ganglia are located close to the effector structures. This makes the postganglionic nerve fibers shorter. Examples are otic ganglion, sphenopalatine ganglion, submaxillary ganglion, ciliary ganglion, among others. Effects of parasympathetic innervation may either be excitatory (involving the development of excitatory postsynaptic potential) or inhibitory. Examples of the inhibitory effects of parasympathetic stimulation include decreased heart rate, the force of cardiac contractility, and blood pressure. Other effects include miosis, increased gastrointestinal motility and secretion, peripheral vasodilation, increased cutaneous blood flow, penile, and clitoral tumescence among others.

The sympathetic nervous system evolved to enable organisms to cope with the emergency, specifically, increased metabolic demand associated with an emergency, flight, and fight. The sympathetic nervous function is executed by the thoracolumbar outflow of the spinal cord and lower brain areas. Unlike the parasympathetic nervous system, the sympathetic ganglia are situated close to the spinal cord and from where postganglionic fibers originate to innervate effectors. The effect of sympathetic stimulation is mediated through binding with adrenergic receptors (alpha and beta receptors). The effects of sympathetic stimulation are increased heart rate and cardiac contractility, increased blood pressure, mydriasis, decreases gastrointestinal motility and secretion, peripheral vasoconstriction among others. It is interesting to note that many phenomena can elicit sympathetic nervous activity. One of them is stress.

Stress is a perception of physical, psychological, or physico-psychological strains or discomforts. During stress, the discomfort perceived physically, psychologically or physical-psychologically represents deviations from what the human body has recognized as normal. Stress is characterized by a nonspecific physiological response to stressors, stress-inducing phenomena. Stress-inducing phenomena can be defined as environmental insults, such as perturbed levels of light, temperature, chemicals, ambient oxygen, and noise. In terms of rapidity of physiological response to stress, the chemical messenger-mediated sympathetic nervous system including the autonomic-adrenal axis plays a major role. Furthermore, the chemical messenger-mediated sympathetic nervous system is a determinant of nonhormonal effects which are often used as general stress markers, such as change in heart rate, heart rate variability, blood pressure, body temperature, blood glucose, baroreflex sensitivity, skin conductance, among others. The chapter looks at different forms of environmental stressors and their autonomic characterizations.

2. Stress

Stress is a perception of physical, psychological, or physico-psychological strains or discomforts. Stress as a sensation of discomfort may be subjective, that is, qualitative. For instance, during a prolonged examination procedure, a student may report having muscle ache or nausea without any confirmatory screenings. In objective cases, medical assessments are done to ratify the existence of stress [1]. Examples of these include determinations of catecholamine, cortisol, prolactin, heart rate, blood pressure, body temperature, brain rhythm, cardiac rhythm, blood glucose, respiratory rate, peripheral oxygen saturation, urine specific gravity among others. During stress, the discomfort perceived either physically or psychologically or both represents deviations from what the human body has recognized as normal. For example, sudden standing elicits perception of physical and physico-psychological strains because of the sudden diversion of a large amount of blood to structures below the heart including the lower extremities at the detriment of other structures including the brain. Events such as driving a car, teaching a crowd of students, having sexual intercourse, exposure to high altitude or low atmospheric oxygen tension, exposure to noise, or running a 100meter race and nighttime study [2, 3] will result in the perception of physical and physico-psychological discomfort in naïve individuals. It is important to add that a measure of stress is beneficial for physical fitness and health status. This is simply known as eustress. An example of eustress is exercise, any programmed and purposeful physical activity. This invariably means that it is not all physical, psychological, and psychophysical sensations of discomfort that can be imaged as harmful. For instance, in order to be able to run a marathon race, a high $V0_2$ max is needed. Achieving this high $V0_2$ max through training is not unassociated with sensations of discomfort. The term "distress" comfortably describes all sensations of discomfort either physically, psychologically or both that are not beneficial to human health and survival [4].

Furthermore, sudden standing, driving a car for the first time, acute exposure to high altitude or low atmospheric oxygen tension, exposure to unfamiliar noise, or executing new physical tasks like running a 100meter race, clearly indicate that stress is part of man. If stress is part of man, then, are there natural defensive, protective, and modulatory mechanisms that come into play during stress? The human body was designed to respond to stress nonspecifically through two mechanisms and they are hormonal and nonhormonal mechanisms. Hormonal mechanisms operate through the autonomic-adrenal axis, hypothalamo-hypophyseal-adrenal axis, tuberoinfundibular dopamine axis, and autocrine, paracrine, and neurohormonal pathways. The nonhormonal mechanisms are outcomes of hormonal mechanisms. Examples include changes in heart rhythms, blood pressure, skin conductance, blood glucose, body temperature, respiratory rate, peripheral oxygen saturated among others.

Any agents or conditions that orchestrate deviations from what the human body has recognized as normal or homeostatic are known as a stressor. They are stress-inducing situations and events [1]. These agents may be exogenous, chemical, and biological. Information made available by Study.com indicates hyperthermia results in discomfort. In diurnal mammals, secretions of melatonin predicates photic signals. Hence, disruption of photic signal and photic stress impairs melatonin secretion resulting in circadian desynchronization, impaired glucose homeostasis [2, 3], weight gain, diabetes mellitus, sexual disorder, and breast cancer [3]. Thermal changes, unfamiliar noise, job-related mental and physical exertions [4], job loss, increase in expected work output, overcrowding, marital problem, irregular lighting, bereavement, barotrauma, traveling, exposure to low atmospheric oxygen, economic hardship, heightened family needs and demands, drug and chemical use, and abuse are stressors. Hence, they demonstrate a huge tendency of triggering certain effects. The effects produced depend on the type of stressor. For instance, metabolic perturbation and cytokine production can be produced by chemical stressors. Intake of tobacco brings about an increase in metabolic rate causing tachycardia, palpitation, tachypnea among others. Alcohol induces secretions of cytokines such as dopamine, endorphin, and serotonin secretion resulting in motivation and addiction. Direct and indirect trauma on the body tissues such as epithelium, muscles, connective tissues, and neural structures and pain-mediating cytokines such as substance P and glutamate are results of both physical and chemical stressors. Both physical and chemical stressors may activate hormonal mechanisms through the autonomic-adrenal axis, hypothalamic hypophyseal-adrenal axis, tuberoinfundibular dopamine axis, autocrine, paracrine, and neurocrine pathways resulting in the release of norepinephrine, dopamine, cortisol, prolactin, prostaglandins interleukin-6, leptin, dehydroepiandrosterone, salivary alpha-amylase, alpha tumor necrosis factor, ghrelin, and growth hormone. As widely believed, human existence has always revolved around not only the surroundings but also the conditions of the environment (Psychology second course) [5].

3. Effect of photic stress on autonomic nervous system

Indiscriminate and prolonged exposure to light has a great deal of impact on the autonomic nervous system. Yasushi *et al.*, [6] examined the effect of exposure to bright light (5000 lux) on sympathetic nervous activity for a period of 20 minutes. The authors reported that exposure increased heart rate while no change was noticed in blood pressure. Muscle sympathetic nerve activity was found to increase after exposure. Sleep is an autonomic phenomenon, Katsanis *et al.*, [7] reported that exposure to color light stimulation shortened time for sleep onset and influence heart rate variability in healthy humans. Specifically, color light preference was found to lower frequency/high-frequency ratio. The study simply indicated that color light preference induces parasympathetic nervous stimulation leading to enhancement of sleep.

To demonstrate the role of dynamic lighting system after fatigue, Maietta *et al.*, [8] utilized a self-reported questionnaire and heart rate variability to measure the autonomic nervous system. There was a rise in parasympathetic nervous mechanisms in the dynamic lighting system within 25 minutes. Heart rate variability means interbeat time interval. Usually, heart rate variability is inversely related to heart rate. This implies the higher the heart rate variability, the small the heart rate. Schaefer and Kraft [9] studied the impacts of light stress courtesy of illumination with a colored fluorescent light on autonomic regulation and heart rate variability. He studies the exposure of 12 healthy individuals to red, blue, and green light (700 lux) for 10 minutes. With a green light, there was a reduction in high-frequency components; thus indicating that illumination with colored light can affect heart rate variability.

Glucose metabolism is influenced by the autonomic nervous system. The islet, the birthplace of insulin and glucagon. is innervated by autonomic nerve fibers. Target tissues, such as the liver, muscles, and adipose tissues, express receptors for these hormones. The expression of these receptors is necessary for the hormone to exert its effects. Hormone resistance or hormone tolerance occurs when there is a reduction in the levels of expressed hormone receptors. Sympathetic stimulation of islet is widely known to increase glucagon secretion. Usually, glucagon, when secreted, is in immature form (preproglucagon) which will become activated through intracellular mechanisms. On the other hand, sympathetic stimulation brings about the breakdown of hepatic glycogen and gluconeogenesis. Exposure to light at night was found to increase glucose intolerance in rats. Specifically, the green light was shown to exert more glucose intolerance. However, blue and red lights did influence glucose intolerance. At 50 and 150 lx, greater glucose responses were produced than 5–20 lx [10].

The study results clearly showed that the intensity of light exposure affects glucose homeostasis.

When C57Bl/6 J mice were placed on constant light in a bid to understand how constant length–induced circadian rhythm abnormality affects energy metabolism and insulin sensitivity, Coomans *et al.,* [11] showed a pronto decline in the amplitude of circadian rhythm. Energy expenditure fell by −13% and food intake rose by +26% culminating in pronto weight gain. Energy metabolisms, insulin sensitivity, and endocrine functions of the islet are autonomically controlled processes. Parasympathetic stimulation of the pancreatic islet through m4 receptors results in a CAMP-mediated increase in the secretion of preproinsulin, which will then be converted to insulin. In the quantification of insulin sensitivity, it is not only insulin secretion that receives consideration. Insulin sensitivity which examines the responsiveness of target tissues to insulin is essential. Insulin sensitivity involves both the reception of insulin by target tissue and associated transduction.

Similarly, a study by Abulmeaty *et al.,* [12] investigated the impact of prolonged continuous exposure on energy homeostasis as well as adropin expression, RORα, and Rev-erb-α nuclear receptors in 32 rats over a period of 3 months. Continuous light exposure was found to raise total energy expenditure. Lowered respiratory quotient and Rev-erb-α hepatic and renal nuclear receptors coupled with elevation in RORα and plasma levels of adropin and expressions. The study indicated the possible involvement of adropin, RORα, and plasma levels of adropin in energy homeostasis. The parasympathetic nervous system is a component of the autonomic nervous system concerned with energy restoration and homeostasis.

Melatonin secretion is an autonomically dependent process. The pineal-innervating postganglionic fibers from the superior cervical ganglion are known to secrete norepinephrine. Elsaid and Fahim [13] showed that maternal exposure to excess artificial light using female rabbits orchestrated a decline in maternal melatonin. In addition, epidermal vacuolation, swollen mitochondria, and shrunken indented nuclei were gotten. As far as the fetus was concerned, reduced thickness of the epidermis, decreased hair follicle number, raised collagen surface area, suppressed proliferating cell nuclear antigen-positive cells were gotten. It is possible that excess light-induced suppression of melatonin is responsible for maternal and fetal histological changes in the study. Furthermore, light also influences body temperature and the reproductive cycle. Specifically, in rodents' study, exposure to light stress has been shown to lengthen the estrous cycle and estrous cycle ratio [2, 3, 14] and impairs body temperature rhythm [15]. Light exposure may also influence the activities of antioxidant enzymes. At least, a study has shown that ovarian glutathione peroxidase increased in light-deprived rats [16].

Sleep is another autonomically controlled mechanism. Sympathetic nervous stimulation through activation of alpha or beta receptors results in increased cortical activities, thereby promoting alertness. On the other hand, the parasympathetic nervous system elicits a reduction in cortical activities, eliciting rest and sleep. Stenvers *et al.,* [17] reported a novel means of creating circadian desynchronization in rats. In their study, they exposed rats to 12-hour of light with the intensity of 150–200 lux and natural 12 hours of dark or 12-hour light (150–200 lux) and 12-hour dark (5 lux). The amplitude of REM and NREM sleep rhythms was attenuated with 12 hour-light (150–200 lux) and 12-hour dark (5 lux). Although 12-hour light (150–200 lux) and 12 hour-dark (5 lux) did not affect glucose tolerance and body weight, there was a reduction in suprachiasmatic expression of the clock gene and desynchronized locomotive activity.

Ishida *et al.*, [18] showed that exposure to light orchestrated the expression of the adrenal gland gene and the underlying mechanism involved suprachiasmatic nucleus-sympathetic nervous system connection. One of the most extensively documented suprachiasmatic nucleus-sympathetic nervous system connections is the retinohypothalamic-pineal gland pathway, through which ambient light passes through intrinsically photosensitive mesopic retinal ganglionic cells to the suprachiasmatic nucleus of the anterior hypothalamus and from there to the pineal gland via superior cervical ganglion. Before the advent of the electric bulb in 1860, human functions were governed by the natural light/dark cycle that consisted of 12 hours of daylight and 12 hours of dark. Cailotto *et al.*, [19] reported that exposure to nocturnal light exerted a rapid effect on peripheral clock gene expression with autonomic hepatic innervation being found to be essential for the photic signals from the suprachiasmatic nucleus.

Kalsbeek *et al.*, [20] identified gamma aminobutyric acid (GABA) as the endogenous mechanism involved in the transmission of photic signals from the suprachiasmatic nucleus to the pineal gland. In the study, the authors showed the role of the chemical messenger by administering GABA antagonist to some parts of the suprachiasmatic nucleus. The presence of GABAergic projection between ventrolateral and dorsolateral parts of the suprachiasmatic nucleus has also previously been reported.

4. Effect of thermal stress on autonomic nervous system

Exposure to thermal stress has a profound effect on the autonomic nervous system. Niimi *et al.*, [21], in a study, examined the effect of heat stress on the sympathetic nervous system in human subjects on the sympathetic nervous system. Nine subjects were deployed for the study and they were exposed to heat stress through the rise in environmental temperature gradually from 290°C to 400°C. Muscle sympathetic nerve activity, tympanic temperature, plasma arginine vasopressin, cutaneous blood flow, heart rate, blood pressure, cardiac output, and mean arterial blood pressure were monitored. The rise in muscle sympathetic nerve activity, tympanic temperature, cardiac output, mean arterial blood pressure, and cutaneous blood flow were recorded when heat stress was applied. Crandall *et al.*, [22] exposed seen subjects to whole-body heating through the infusion of 15 ml/kg of warm water. Skin blood flow, heart rate, arterial blood pressure, muscle sympathetic nerve activity, and sublingual temperature were measured. The authors observed that thermal exposure raised muscle sympathetic nerve activity but central venous blood and mean arterial blood pressure were unaffected. However, mean arterial pressure was found to increase when the effect of arterial baroreceptor loading on muscle sympathetic nerve activity during heat stress was conducted. Low *et al.*, [23] examined the possibility that a shift from moderate to severe thermal stress would raise muscle sympathetic nerve activity and sympathetic nerve activity continuously. Thirteen subjects were made to undergo passive thermal stress that raised body temperature by 1.3°C. There was a rise in mean cutaneous temperature. Mean body temperature, heart rate, and skin vascular conductance by thermal stress exposure. There was also a transient increase in muscle sympathetic nerve activity with thermal stress. Cui *et al.*, [24] investigated the role of thermal stress in the alleviation of muscle sympathetic nerve activity during gravitational stress. The authors found that there was a rise in sublingual temperature, sweat rate, cutaneous blood flow, heart rate, and muscle sympathetic nerve activity during whole-body heating. However, the rise in muscle sympathetic nerve activity which occurred in response to gravitational stress was not assuaged by thermal stress in human subjects. Cutaneous heating depressed the high-frequency component of heart rate variability but raised the ratio of low-to-high frequency component and mayer waves. However, there was no increase in baroreflex sensitivity with cutaneous heating [25]. Kaho *et al.*, [26] investigated how thermal stress affects heart rate variability in freely moving ruminant animals. They measured autonomic nervous system markers such as heart rate variability, temperature-humidity index, among others. Heart rate variability was reported to decline with temperature-humidity index. Their study showed that high thermal stress may affect the autonomic equilibrium of ruminant organisms by inducing the sympathetic nervous system nonlinearly. Wilson and Ray, [27] reported that whole-body heating raised heart rate, internal temperature, and muscle sympathetic nerve activity but mean arterial blood pressure was unaffected.

5. Effect of chemical stress on autonomic nervous system

Chemical stressors such as exogenous chemicals have an influence on the autonomic nervous system. Geraldes *et al.*, [28] reported increased chemosensitivity as well as baroreceptor reflex disruption, sympathetic hyper-excitation, increased heart rate, and high blood pressure. Shvachiy *et al.*, [29] investigated the impact of intermittent exposure to lead on autonomically controlled body functions such as cardiovascular and respiratory functions. High blood pressure, impaired baroreflex sensitivity and tachypnea were recorded following intermittent exposure to lead. In addition to the fact that chemical stressors suppressed parasympathetic actions in infancy and childhood, the effect may be scourged even during intrauterine life. Jurczyk *et al.*, [30] in their review showed that prenatal exposure to alcohol decreased heart rate variability. Prolonged alcohol exposure has been linked with an increased risk of heart disease [31]. Decreased heart rate variability implies a high heart rate. Yu *et al.*, [32] intended to examine the association between blood lead and autonomic nervous activity (heart rate variability and median nerve conduction velocity in 328 adult males aged 28 years. The authors found no association between blood level concentration and autonomic nervous activity. Liao *et al.*, [33] reported that there was an association between exposure to airborne heavy metals and autonomic dysfunction in 82 young adults. Other airborne chemical stressors which have been reported to impair autonomic nervous activity, raising systolic blood pressure and diastolic blood pressure and disturbing other autonomically controlled structures include dust [34], cadmium [35], aluminum [36], lead [37], and oil spillage [38].

6. Effect of hypoxic stress on autonomic nervous system

The environment is the primary source of oxygen in the body. Hypoxic stress represents the sensation of discomfort associated with exposure to the low partial pressure of oxygen. Exposure to high altitudes (3000–4000 m) has been reported to lead to depression of heart rate variability power spectra [39]. Exposure to hypoxia and hypercapnia has been reported to raise sympathetic burst frequency. However, hypoxia was found to produce a long-lasting effect on sympathetic activation in human subjects than exposure to hypercapnia [40]. Excessive hypoxia also causes disruption of peripheral chemoreceptors and hyperviscosity. Hyperviscosity leads to increased pulmonary blood pressure, right cardiac failure, inadequate brain perfusion, and death [41].

7. Effect of noise stress on autonomic nervous system

Unlike note, noise is produced when sound waves are emitted from multiple sources. Noise has been reported to impact autonomic nervous activity. Goyal *et al.*, [42] exposed 200 adult males to noise intensity of greater than 80 dB for about 6 months. Noise exposure increased heart rate and systolic blood pressure. Idrobo-Ávila *et al.*, [43] in their systematic review noted that noise exposure increased heart rate and blood pressure and caused alteration in both high frequency and low frequency/high-frequency ratio. Chen *et al.*, [44] examined the connection between occupational exposure to noise and high blood pressure using 1390 workers exposed to occupational noise. The authors reported that noise exposure caused an increase in systolic blood pressure and diastolic blood pressure. According to the study, the prevalence of high blood pressure was found as 17.8% in those who were exposed to noise and 9% in those who were not exposed to noise. Hey also claimed that there existed a stronger regression coefficient between diastolic blood pressure and noise exposure. Noise intensity, cumulative noise exposure, and years of exposure were related to high blood pressure risk. de Souza *et al.*, [45] claimed that exposure to noise was associated independently with high blood pressure at 75–85 dB. The study recruited 1729 subjects who were petrochemical workers. The authors also found that age, body mass index, and gender were associated with high blood pressure independently. A study by Brahem *et al.*, [46] was conducted on 120 electricity workers who were exposed to noise. Prevalence of high blood pressure was found in workers who were exposed to noise. Systolic blood pressure and diastolic blood pressure were higher in workers who were exposed to noise. Kuang *et al.*, [47] showed that increasing years of exposure to occupational noise increased systolic blood pressure and diastolic blood pressure.

8. Summary

Environmental stressors such as ambient threshold changes in light, heat, noise atmospheric oxygen, and chemical phenomena have a great impact on the autonomic nervous system causing impairment in general stress markers. Environmental stressors orchestrated an increase in systolic blood pressure, diastolic blood pressure, mean arterial blood pressure and disrupted baroreflex sensitivity in both males and females. Derangement in body temperature and blood glucose and suppression of melatonin were also consequences of environmental stressors in both human beings and animals. Future studies will examine in detail the molecular mechanisms that underlie environmental stressor-induced changes in general stress markers.

Author details

Mayowa Adeniyi
Stress and Physical Exertion Unit, Department of Physiology, Edo State University Uzairue, Iyamho, Edo State, Nigeria

*Address all correspondence to: 7jimade@gmail.com

IntechOpen

© 2022 The Author(s). Licensee IntechOpen. This chapter is distributed under the terms of the Creative Commons Attribution License (http://creativecommons.org/licenses/by/3.0), which permits unrestricted use, distribution, and reproduction in any medium, provided the original work is properly cited.

References

[1] Shawi A, Abdullateef AN, Khedher MA, Rejab MS, Khaleel RN. Assessing stress among medical students in Anbar governorate, Iraq: a cross-sectional study. The Pan African Medical Journal. 2018;**31**:96. DOI: 10.11604/pamj.2018.31.96.16737

[2] Adeniyi MJ, Fabunmi O, Okojie AK, Olorunnisola OL, Odetola AO, Olaniyan OT, et al. Impact of night study frequency on sleep pattern, anthropometrical indices and peripheral oxygen saturation in age-matched nigerian female students prior to semester examination. International Journal of Biomedical Sciences. 2020a;**16**(3):37-42

[3] Adeniyi MJ, Agoreyo FO, Olorunnisola OL, Olaniyan OT, Seriki SA, Ozolua PO, et al. Photo-pollution disrupts reproductive homeostasis in female rats: The duration-dependent role of selenium administrations. The Chinese Journal of Physiology. 2020b;**63**:235-243

[4] Ulrich-Lai YM, Fulton S, Wilson M, Petrovich G, Rinaman L. Stress exposure, food intake and emotional state. Stress. 2015;**18**(4):381-399. DOI: 10.3109/10253890.2015.1062981. Epub 2015 Aug 13

[5] Psychology Second Course 105-115. 222en23. Available from: https://nios.ac.in/media/documents/secpsycour/English/Chapter-23.pdf [Last accessed: October 22, 2021]

[6] Saito Y, Shimizu T, Takahashi Y, Mishima K, Takahashi K-i, Ogawa Y, et al. Effect of bright light exposure on muscle sympathetic nerve activity in human. Neuroscience Letters. 1996;**219**(2):135-137

[7] Lee S, Kim D. Effect of Color Light Stimulation Using LED on Sleep Induction Time. Journal of Healthcare Engineering. 2017;**2017**:6030268. DOI: 10.1155/2017/6030268. Epub 2017 May 11

[8] Maietta S, Chen C-Y, Wu P-J, Hsiao Y-J, Tai Y-W. Autonomic nervous system under dynamic lighting environment during a short rest. Changes in Humans&#x0027. 2021;**2021**:6697701. DOI: 10.1155/2021/6697701

[9] Schaefer A, Kratky KW. The effect of colored illumination on heart rate variability. Forschende Komplementärmedizin. 2006;**13**:167-173

[10] Opperhuizen AL, Stenvers DJ, Jansen RD, Foppen E, Fliers E, Kalsbeek A. Light at night acutely impairs glucose tolerance in a time-, intensity- and wavelength-dependent manner in rats. Diabetologia. 2017;**60**:1333-1343. DOI: 10.1007/s00125-017-4262-y

[11] Coomans CP, van den Berg SA, Houben T, van Klinken JB, van den Berg R, Pronk AC, et al. Detrimental effects of constant light exposure and high-fat diet on circadian energy metabolism and insulin sensitivity. The FASEB Journal. 2013;**27**(4):1721-1732. DOI: 10.1096/fj.12-210898. Epub 2013 Jan 9

[12] Abulmeaty MMA, Almajwal AM, Alnumair KS, Razak S, Hasan MM, Fawzy A, et al. Effect of long-term continuous light exposure and western diet on adropin expression, lipid metabolism, and energy homeostasis in rats. Biology (Basel). 2021;**10**(5):413. DOI: 10.3390/biology10050413

[13] Elsaid AG, Faheem NM. Impact of constant light exposure duringpregnancy

on skin of neonatal New Zealand rabbits: structural and ultrastructural study. Brazilian Journal of Medical and Biological Research. 2021;**54**(6):e10722. DOI: 10.1590/1414-431X202010722

[14] Adeniyi MJ, Agoreyo FO. Estrous cycle ratio as a reproductive index in the rats. American Journal of Biomedical Science and Research. 2019;**4**(2):100-103

[15] Adeniyi MJ, Agoreyo FO. Duration-related modulation of body temperature rhythm and reproductive cycle in rats by photoperiodic perturbation. DRJHP. 2020;**8**(1):1-6

[16] Agoreyo FO, Adeniyi MJ. Pattern of estrous cycle and ovarian antiperoxidative activity in light deprived sprague-dawley rats treated with sodium. Journal of Medicinal Research and Biological Studies. 2018;**1**(1):103

[17] Stenvers DJ, van Dorp R, Foppen E, Mendoza J, Opperhuizen AL, Fliers E, et al. Dim light at night disturbs the daily sleep-wake cycle in the rat. Scientific Reports. 2016;**20**(6):35662. DOI: 10.1038/srep35662

[18] Ishida A, Mutoh T, Ueyama T, Bando H, Masubuchi S, Nakahara D, et al. Light activates the adrenal gland: timing of gene expression and glucocorticoid release. Cell Metabolism. 2005;**2**(5):297-307. DOI: 10.1016/j.cmet.2005.09.009

[19] Cailotto C, Lei J, van der Vliet J, van Heijningen C, van Eden CG, Kalsbeek A, et al. Effects of nocturnal light on (clock) gene expression in peripheral organs: a role for the autonomic innervation of the liver. PLoS One. 2009;**4**(5):e5650. DOI: 10.1371/journal.pone.0005650

[20] Kalsbeek A, Cutrera RA, Van Heerikhuize JJ, Van Der Vliet J, Buijs RM. GABA release from suprachiasmatic nucleus terminals is necessary for the light-induced inhibition of nocturnal melatonin release in the rat. Neuroscience. 1999;**91**(2):453-461. DOI: 10.1016/s0306-4522(98)00635-6

[21] Niimi Y, Matsukawa T, Sugiyama Y, Shamsuzzaman AS, Ito H, Sobue G, et al. Effect of heat stress on muscle sympathetic nerve activity in humans. Journal of the Autonomic Nervous System. 1997;**63**(1-2):61-67. DOI: 10.1016/s0165-1838(96)00134-8

[22] Crandall CG, Etzel RA, Farr DB. Cardiopulmonary baroreceptor control of muscle sympathetic nerve activity in heat-stressed humans. The American Journal of Physiology. 1999;**277** (6 Pt 2):H2348-H2352. DOI: 10.1152/ajpheart.1999.277.6.h2348

[23] Low DA, Keller DM, Wingo JE, Brothers RM, Crandall CG. Sympathetic nerve activity and whole body heat stress in humans. Journal of Applied Physiology. 1985;**111**(5):1329-1334. DOI: 10.1152/japplphysiol.00498.2011 Epub 2011 Aug 25

[24] Cui J, Wilson TE, Crandall CG. Muscle sympathetic nerve activity during lower body negative pressure is accentuated in heat-stressed humans. Journal of Applied Physiology. 1985;**96**(6):2103-2108. DOI: 10.1152/japplphysiol.00717.2003

[25] Geraldes V, Carvalho M, Goncalves-Rosa N, Tavares C, Laranjo S, Rocha I. Lead toxicity promotes autonomic dysfunction with increased chemoreceptor sensitivity. Neurotoxicology. 2016;**54**:170-177. DOI: 10.1016/j.neuro.2016.04.016 Epub 2016 Apr 28

[26] Kitajima K, Oishi K, Masafumi M, Hirok A, Akira S, Yudai Y, et al. Effects of heat stress on heart rate variability in

free-moving sheep and goats assessed with correction for physical activityof muscle sympathetic nerve activity in heat-stressed humans. The American Journal of Physiology. 1999;**277**(6 Pt 2):H2348-H2352. DOI: 10.1152/ajpheart.1999.277.6.h2348

[27] Wilson TE, Ray CA. Effect of thermal stress on the vestibulosympathetic reflexes in humans. Journal of Applied Physiology. 2004;**97**(4):1367-1370

[28] Geraldes V, Carvalho M, Goncalves Rosa N, Tavares C, Laranjo S, Rocha I. Lead toxicity promotes autonomic dysfunction with increased chemoreceptor sensitivity. Neuro-toxicology 2016;54:170-177. DOI: 10.1016/j.neuro.2016.04.016. Epub 2016 Apr 28

[29] Shvachiy L, Geraldes V, Amaro Leal Â, Rocha I. Intermittent low-level lead exposure provokes anxiety, hypertension, autonomic dysfunction and neuroinflammation. Neurotoxicology. 2018;**69**:307-319. DOI: 10.1016/j.neuro.2018.08.001 Epub 2018 Aug 8

[30] Jurczyk M, Dyląg KA, Skowron K, Gil K. Prenatal alcohol exposure and autonomic nervous system dysfunction: A review article. Folia Medica Cracoviensia. 2019;**59**(3):15-21. DOI: 10.24425/fmc.2019.131132

[31] Ige SF, Adeniyi MJ, Adeniyi AM, Ibrahim SD. Hematological and immunological characterizations of experimental ulcerative colitis rats: The impact of alcohol consumption pattern. International. Journal of Biomedical Science. 2021;**17**(2):6-14

[32] Yu CG, Wei FF, Yang WY, Zhang ZY, Mujaj B, Thijs L, et al. Heart rate variability and peripheral nerve conduction velocity in relation to blood lead in newly hired lead workers. Occupational and Environmental Medicine. 2019;**76**(6):382-388. DOI: 10.1136/oemed-2018-105379 Epub 2019 Mar 30

[33] Liao YH, Chen WL, Wang CC, Lai CH. Associations between personal exposure to metals in fine particulate matter and autonomic nervous system dysfunction among healthy adults. Aerosol and Air Quality Research. 2020;**20**:1842-1849. DOI: 10.4209/aaqr.2020.04.0156

[34] Oni TJ, Adeniyi MJ. Postural difference in expiratory rate among female sanitary workers and its relationship with blood pressure and anthropometric indices. Biomedical Journal of Scientific & Technical Research. 2017;**1**(2):311-315

[35] Ige SF, Adeniyi MJ, Ademilua OB, Fatola AO, Adeyemi IA. Allium cepa remediates oxidative stress-mediated hepatic DNA damage in cadmium-exposed rats through enhanced p53 expression and inhibition of bcl2. International Journal of Biomedical Sciences. 2020;**16**(2):11-17

[36] Ige SF, Adeniyi MJ, Iyalla GO. Allium cepa mitigates aluminum chloride induced hepatotoxicity in male wistar rats. Journal of Biomedical Science. 2017;**6**(4):27

[37] Ige SF, Adeniyi MJ, Joanna AO, Oluwaseyi DA. Allium Cepa juice prevented oxidative stress-mediated metabolic disorder following chronic lead acetate exposure in male rats. Albanian Journal of Medical and Health Science. 2019;**1**(40):11

[38] Adeniyi MJ, Oni TJ. Water contamination in Nigeria and body defense issues. Research Chronicler. 2016;**3**(3):11-21

[39] Zhang D, She J, Zhang Z, et al. Effects of acute hypoxia on heart rate variability, sample entropy and cardiorespiratory phase synchronization. Biomedical Engineering Online. 2014;**13**:73. DOI: 10.1186/1475-925X-13-73

[40] Xie A, Skatrud JB, Puleo DS, Morgan BJ. Exposure to hypoxia produces long-lasting sympathetic activation in humans. Journal of Applied Physoplpgy. 2001. DOI: 10.1152/jappl.2001.91.4.1555

[41] Hainsworth R, Drinkhill MJ, Rivera-Chira M. The autonomic nervous system at high altitude. Clinical Autonomic Research. 2007;**17**:13-19

[42] Goyal S, Gupta V, Walia L. Effect of noise stress on autonomic function tests. Noise & Health. 2010;**12**(48):182-186. DOI: 10.4103/1463-1741.64976

[43] Idrobo-Ávila EH, Loaiza-Correa H, van Noorden L, Muñoz-Bolaños FG, Vargas-Cañas R. Different types of sounds and their relationship with the electrocardiographic signals and the cardiovascular system - review. Frontiers in Physiology. 2018;**22**(9):525. DOI: 10.3389/fphys.2018.00525

[44] Chen S, Ni Y, Zhang L, Kong L, Lu L, Yang Z, et al. Noise exposure in occupational setting associated with elevated blood pressure in China. BMC Public Health. 2017;**17**(1):107. DOI: 10.1186/s12889-017-4050-0

[45] de Souza TC, Périssé AR, Moura M. Noise exposure and hypertension: investigation of a silent relationship. BMC Public Health. 2015;**3**(15):328. DOI: 10.1186/s12889-015-1671-z

[46] Brahem A, Riahi S, Chouchane A, Kacem I, Maalel OE, Maoua M, et al. Impact du bruit professionnel sur le développement de l'hypertension artérielle : enquête réalisée au sein d'une centrale de production d'électricité et de gaz en Tunisie [Impact of occupational noise in the development of arterial hypertension: A survey carried out in a company of electricity production]. Annales de Cardiologie et d'Angéiologie (Paris). 2019;**68**(3):168-174. French. DOI: 10.1016/j.ancard. 2018.10.008

[47] Kuang D, Yu YY, Tu C. Bilateral high frequency hearing loss is associated with elevated blood pressure and increased hypertension risk in occupational noise exposed workers. PLoS One. 2019;**14**(9):e0222135. DOI: 10.1371/journal.pone.0222135

Section 3

Special Interest Topics

Chapter 5

General Anesthesia and Autonomic Nervous System: Control and Management in Neurosurgery

Irina Alexandrovna Savvina, Anna Olegovna Petrova and Yulia Mikhailovna Zabrodskaya

Abstract

The chapter is devoted to the control and management of the autonomic nervous system during general anesthesia in neurosurgery. The brainstem and supratentorial cerebral centers of autonomic regulation are the most important structures for control and management during general anesthesia using pharmacological defense with α2-adrenergic agonists and opioid analgesics. We discuss the questions of the depth of anesthesia (BIS-monitoring) and antinociceptive defense, variability of heart rate (variational cardiointervalometry), hemodynamic monitoring during neurosurgical operation, intraoperative thermometry, the meaning of trigeminocardiac reflex and its classification in neurosurgery, perioperative events causing autonomic distress syndrome development and methods of its prophylaxis and treatment, pathomorphological signs of vegetative distress syndrome. Control of the neuromuscular block and photoplethysmography assessment of perfusion index (PI) as methods of the adequacy of general anesthesia and neurovegetative stability.

Keywords: general anesthesia, premedication, autonomic nervous system, neurosurgery, brainstem autonomic centers, depth of anesthesia, trigeminocardiac reflex, variability of heart rate, autonomic distress syndrome

1. Introduction

Scientific and clinical interest in the problem of control and management of the autonomic nervous system in various fields of surgery and anesthesiology is due, in our opinion, to the relationship of the initial vegetative status of the patient (tone, reactivity of the ANS) and anesthesia techniques with the peculiarities and complications of the intra- and postoperative period; the development of cardiovascular (cardiac arrhythmias, hemodynamic instability, arterial hypertension), respiratory complications during surgery and in the early postoperative period, the occurrence of postoperative nausea and vomiting (PONV), postoperative delirium (POD), the severity of pain syndrome.

In neurosurgery, the relevance of this problem is due to the participation of autonomous control mechanisms in the autoregulation of cerebral circulation and cerebral vascular tone, which directly affects the intraoperative state of the brain, perfusion pressure of the brain, stability of its volume, and compliance—the most important characteristics reflecting the adequacy of anesthetic provision.

The literature describes prognostic predictors of complications, such as postoperative delirium, with the sympathetic pattern of the autonomic nervous system in otolaryngology surgery [1, 2], the formation of neuropathic pain and postoperative nausea and vomiting with the predominance of parasympathetic influences, etc.

Thus, it is extremely important to take into account the initial version of the tone of the autonomic nervous system. Monitoring and managing the balance and reactivity of the links of the autonomic nervous system during the preparation of the patient for surgery and during the operation and anesthesia itself will allow to obtain a result in the form of reducing the risks of perioperative complications.

2. The brain and the autonomic nervous system

2.1 The structure of the sympathetic and parasympathetic links of the autonomic nervous system

The leading role in maintaining the constancy of the internal environment of the body is played by the department of the nervous system, which regulates the activity of internal organs, glands of internal and external secretion, blood, and lymphatic vessels—the autonomic nervous system.

Autonomic neurons are located mainly in the spinal cord—sympathetic in the thoracic region, parasympathetic in the sacral region (**Figure 1A** and **B**).

2.1.1 Segmental division of the autonomic nervous system

Segmental parts are also embedded in the brain stem—the nuclear apparatus of the vagus nerve; the vegetative nucleus of the VII nerve, the fibers to the sublingual and submandibular glands and the vessels of the meninges; the vegetative nucleus of the IX nerve, from which the tympanic nerve begins, going to the parotid gland, and the

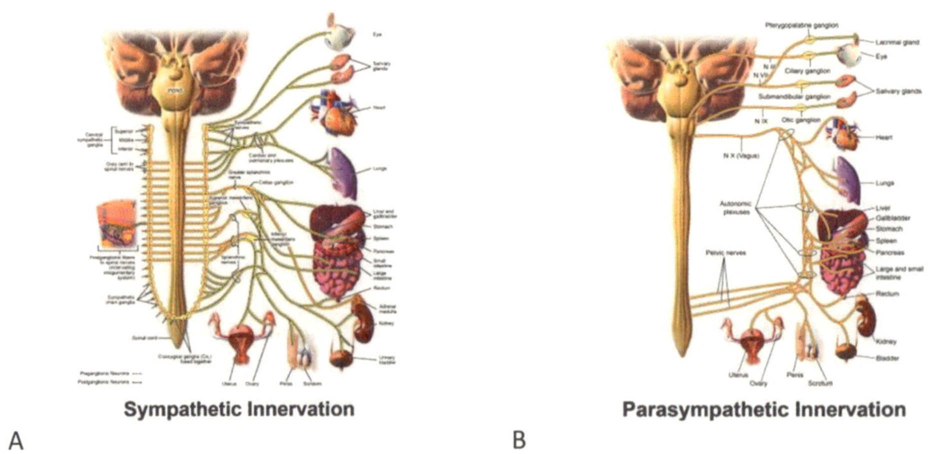

Figure 1.
The structure of the sympathetic (A) and parasympathetic (B) links of the autonomic nervous system.

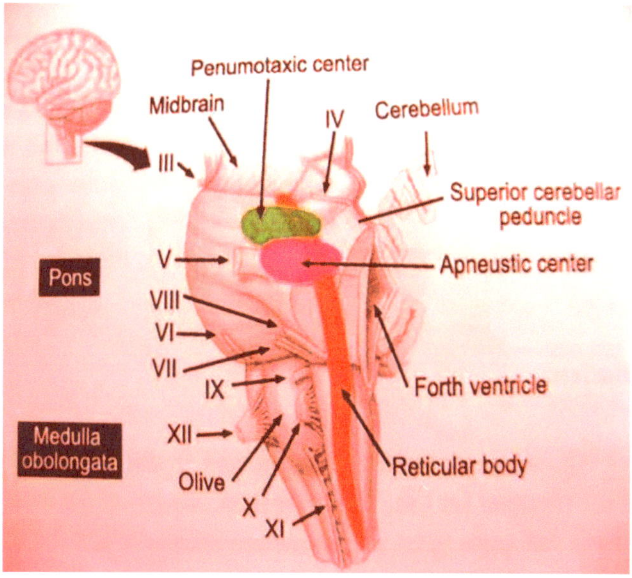

Figure 2.
Brainstem autonomic nervous centers [4].

vegetative nucleus of the oculomotor nerve (Yakubovich-Edinger-Westphal nucleus), the fibers of which are involved in the regulation of pupil size (**Figure 2**) [3, 4].

The number of neurons included in segmental devices exceeds the number of neurons in the brain [5]. Stem nuclear formations are homologs of the lateral horns of the spinal cord, as well as motor and sensory nuclei of the brain stem are homologs of the anterior and posterior horns.

2.1.2 Suprasegmental vegetative nervous system

This integrative system combines the reticular formation of the brain stem, hypothalamus, thalamus, amygdala, hippocampus, septum, together with their connecting paths, which form a functional system, called the limbic-reticular complex (**Figure 3**) [4].

The limbic system provides the integration of somatic and vegetative nervous system—regulation of the autonomic, hormonal functions that provide various forms of activity, including in conditions of general anesthesia and surgical exposure [6, 7].

2.2 Cerebral blood flow and autonomic nervous system

Both parts of the ANS—sympathetic and parasympathetic—are actively involved in the regulation of cerebral vascular tone and blood supply to the brain [8]. The leading role in the sympathetic innervation of cerebral vessels is played by the upper cervical and stellate ganglia. The proximal 2/3 of the basilar artery and vertebral arteries are innervated by the superior cervical ganglion. The anterior, middle, and posterior arteries receive nerve fibers from ganglion cells of the trigeminal ganglion [9].

The sympathetic nervous system participates in protecting the microcirculation of the brain from hemodynamic overloads and thereby preserves the blood-brain barrier and protects brain tissue during significant increases in systemic blood pressure in extreme situations [5, 8, 10]. It is assumed that the trigger for autoregulation

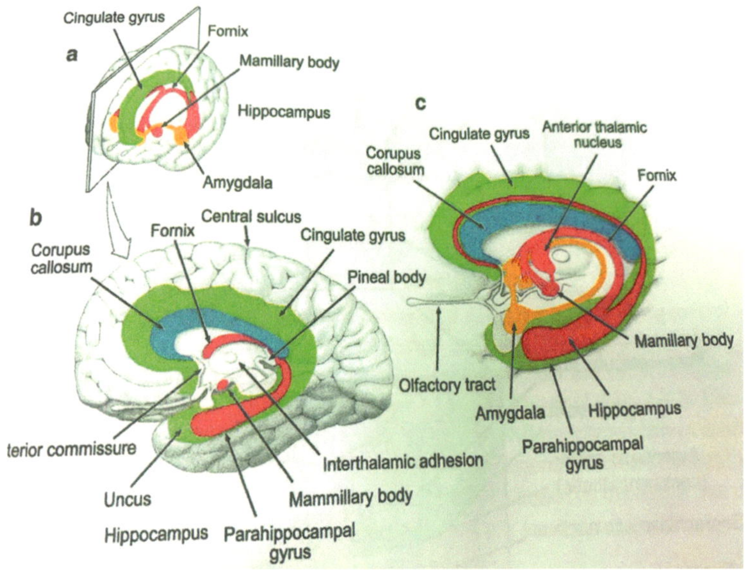

Figure 3.
Limbic system of the brain. Sagittal section through the brain hemispheres (a). Medial view from the right hemisphere of the brain (b). Enlarged image of the limbic system (c) [4].

of excessive narrowing of the cerebral arteries is reflexes from the baroreceptors of the aorta, carotid sinuses, and dura mater, which are realized through the sympathetic nervous system [11]. In response to the stimulation of these receptors, central impulses arise, traveling along the sensitive fibers of the IX and X pairs of cranial nerves to the nucleus of a single pathway located in the medulla oblongata. After processing these signals via efferent pathways, information reaches the executive organs—the heart, blood vessels, kidneys, adrenal glands, and also through the participation of neuroregulatory systems, an integrative response of the brain is triggered in the mechanisms of autoregulation of cerebral circulation [8].

Stimulation of parasympathetic nerves increases cerebral blood flow. The mechanisms of vasodilation of brain vessels during activation of the parasympathetic nervous system are not specified. It is assumed that under the influence of acetylcholine, the content of cyclic guanosine monophosphate (cGMP) increases and the activity of cGMP-dependent protein kinase increases. It is also possible that under the influence of acetylcholine released in synapses, the exchange of potassium and sodium ions changes, and sodium-potassium ATPase is also involved [12]. The large arteries of the base of the brain are innervated by serotonin-containing fibers of sympathetic origin. To date, it has been established that constriction or dilation of blood vessels under the influence of serotonin is caused both by direct action on smooth muscle cells of cerebral vessels, and indirectly through activation of serotonin receptors (HT1, HT2) on perivascular nerve terminals or vascular endothelial cells [13, 14].

2.3 Mechanisms for the implementation of stress reactions with the participation of autonomous regulation centers

The stress reactions are based on activation and tension of the hypothalamic-pituitary-adrenal system and the adrenergic system [15, 16], then an increased synthesis of glucocorticoids and the release of catecholamines in the blood and target organs is triggered [17]. The hippocampus has an inhibitory effect on the neurosecretory system of the hypothalamus and protects it from excessive stress [18]. The hippocampus is also able to inhibit adrenocortical activity and, thus, influence the duration and dynamics of the stress reaction [19, 20]. It is known that emotional stress triggers a powerful stimulation of sympathetic arousals [21, 22], then there is a decrease in the sympathetic and secretory activity of the adrenal glands [23–25], and the body moves to a different metabolic level with the formation of stress resistance [26, 27].

An imbalance of the links of the autonomic nervous system can lead to the development of autonomic distress syndrome in the perioperative period [5, 28, 29].

2.4 Surgical stress and central mechanisms of stress response realization

It is proved that various neurotransmitter systems of the brain are involved in the reactions of the central nervous system to surgical stress—the adrenergic, dopaminergic, cholinergic system since pathological reflexes from the surgical wound are realized through the vagal nuclear complex in the form of vegetovisceral efferent responses [12]. A hypermetabolic state causes [30, 31] the need for tissues for oxygen increases, and the activity of the cardiovascular system increases [15, 32–35]. It is proved that the intensity of reactions of the links of the hypothalamic-pituitary-cortical-adrenal system depends on the type of stress factor and the initial functional state of this system [36, 37]. Arteriole spasm leads to an increase in total vascular resistance, microcirculatory and rheological disorders, the consequence of these pathophysiological changes will be a redistribution of the volume of circulating blood, hypovolemia, tissue and organ ischemia and hypoperfusion, violations of acid-base and water-electrolyte balance, increased peroxidation reactions [38]. Surgical stress causes changes in the permeability of cell membranes, their ultrastructural damage, which will result in a decrease in the functional reserves of organs [36]. The result of surgical stress will be the development of multiple organ dysfunction with the progression of cardiovascular, respiratory failure; impaired liver, kidney, gastrointestinal tract, immunological reactivity, and regulation of the aggregate state of blood in the form of hypercoagulation [29, 39].

3. The role of the autonomic nervous system in limiting stress reactions under conditions of neurosurgical influence

The autonomic nervous system has a modulating effect on compensation mechanisms and their adequacy in response to surgical trauma [40]. The brainstem and supratentorial cerebral centers of autonomic regulation are the most important structures for control and management during general anesthesia using pharmacological defense with α2-adrenergic agonists and opioid analgesics. Daily monitoring of heart rate variability in neurosurgical patients, along with the calculation of the autonomic Kerdo index, in the early postoperative period showed distinct eutonia after removal of a brain tumor under general anesthesia with the opioid analgesic fentanyl in combination with the α2-adrenergic agonist clonidine [41]. Dysfunction of the autonomic nervous system can lead to disruption of adaptation in response to surgical intervention, the development of severe hemodynamic reactions, and complications of the early postoperative period [28, 42–44].

3.1 Assessment of the functional state of the autonomic nervous system in neurosurgical patients

To assess the vegetative status of neurological patients, some authors have proposed generalizing methods [6, 45]. To assess the state of the autonomic nervous system in the perioperative period, some authors used indicators such as the autonomic Kerdo index, a study of daily heart rate variability and their mathematical models, including in patients with intracranial hypertension [46–48].

3.1.1 Assessment of the type of vegetative tone

The Kerdo index is the "gold standard" for assessing the type of vegetative tone. When studying the ratio of diastolic pressure and the number of pulse beats per minute, it was suggested that changes in the ratio of diastolic pressure and the number of pulse beats are associated with shifts in vegetative tone.

The calculation of the vegetative Kerdo index is carried out according to the formula:

$$VI = (1 - \mathrm{D} / HR) \times 100 \tag{1}$$

where VI is the vegetative index, D is the value of diastolic pressure; HR is the heart rate in 1 minute.

Interpretation of the results—complete vegetative equilibrium (eutonia)—VI = 0 – +7; sympathotonia—VI > +7; parasympathotonia—VI < +7 and negative values.

The evaluation of the indicator in dynamics allows us to trace the degree of stress and drug effects on the tone of the ANS.

The interpretation of the calculated values assumes that the minute volume (MV) of the heart with sympathotonia is greater than in a calm state with parasympathotonia. In turn, MV is associated with the compensation of the circulating blood volume (CBV) by peripheral resistance within physiological boundaries. It can be assumed that fluctuations in the minute volume are approximately expressed in terms of pulse rate, and changes in peripheral resistance are expressed through diastolic pressure. This explains the fact that with sympathotonia the pulse rate increases and the diastolic pressure decreases; with parasympathotonia the pulse rate decreases and the diastolic pressure increases. This implies a decrease or increase in the vegetative index toward negative or positive values.

3.1.2 Assessment of vegetative reactivity

The assessment of vegetative reactivity in our study was carried out using the Dagnini-Aschner test and the Cermak-Goering sinocarotide reflex.

3.1.2.1 The ocular reflex (Danyini-Ashner)

The test was carried out as follows—after 15 minutes of lying at rest, the patient's heart rate is calculated for 1 minute (baseline background). Then the pads of the fingers are pressed on both eyeballs until a slight pain appears. After 15–25 seconds, the heart rate is recorded for 20 seconds.

Normally, after a few seconds from the beginning of the pressure, the heart rate slows down by 6–12 beats per 1 minute.

With a normal slowing of the heart rate, normal vegetative reactivity is noted; with a strong slowdown (parasympathetic, vagal reaction)—increased vegetative reactivity; with a weak slowdown—decreased vegetative reactivity; in the absence of slowing—perverted vegetative reactivity (sympathetic reaction).

Due to the different initial heart rates (more or less than 70–72 beats per 1 minute), it is possible to calculate according to the Galu formula:

where HRS is the heart rate in the sample; HRI is the initial heart rate.

The slowing down of the pulse according to the Galu formula is equal to:

$$X = 100 * HRS / HRI \tag{2}$$

The normal value for the ocular reflex is −3.95 ± 3.77.

3.1.2.2 Evaluation of the Cermak - Goering sinocarotid reflex

The technique of the test—after a 15-minute adaptation (rest) in the supine position, the heart rate is calculated in 1 minute—the initial background. Then alternately (after 1.5–2 seconds), the fingers (index and thumb) are pressed on the area of the upper third of the m. sternocleidomastoideus slightly below the angle of the lower jaw until the carotid artery pulsates. It is recommended to start the pressure from the right side since the effect of irritation on the right is stronger than on the left. The pressure should be light, not causing pain, for 15–20 seconds. From the 15th second, the heart rate begins to register for 10–15 seconds. Then the pressure is stopped and the heart rate is calculated in a minute. It is also possible to register the state of the after effect at the 3rd and 5th minutes after the cessation of pressure. Sometimes blood pressure and respiratory rate are recorded.

Interpretation—the values obtained in healthy subjects, that normal vegetative reactivity, are taken as a normal change in heart rate. The normal value of M ± a for the synocarotide reflex is 4.9 ± 2.69.

Values above normal indicate increased vegetative reactivity, that is increased parasympathetic or lack of sympathetic activity, lower—a decrease in vegetative reactivity. An increase in heart rate indicates a perverse reaction.

3.1.2.3 The study of the functions of the segmental part of the autonomic nervous system

The study of the functions of the segmental part of the autonomic nervous system is carried out by conducting an orthostatic test.

The state of the sympathetic efferent pathway is determined according to changes in blood pressure associated with the transition to the vertical position of the patient. The difference in systolic blood pressure is calculated in the supine position and at the 3rd minute after the patient gets up.

Interpretation—a decrease in systolic blood pressure by less than 10 mm Hg is a normal reaction indicating the integrity of efferent vasoconstrictor fibers; a decrease by 11–29 mm Hg is a borderline reaction; a drop by 30 mm Hg and more is a pathological reaction indicating efferent sympathetic insufficiency.

The state of the parasympathetic efferent pathway is determined by measuring the heart rate when getting up. In healthy people, the heart rate increases rapidly when getting up (the maximum figure is noted after the 15th heartbeat) and then decreases after the 30th heartbeat. Normally, the quotient of the division of the first value to the second should be equal to 1.04 or more; 1.01–1.03—borderline result; 1.00—insufficiency of vagal influences on the heart.

3.2 Localization of the brain tumor and vegetative status

The results obtained in the neurosurgical clinic allowed us to conclude that when the tumor was localized in the middle and posterior cranial fossa, there was a predominance of activity of the parasympathetic link of the nervous system [41]. Dysfunction of stem structures in the posterior cranial fossa, irritation of the brain stem due to tumor growth with irritation of the nuclei of the caudal group of cranial nerves, in patients with a tumor of the IV ventricle—nuclei and formations of the rhomboid fossa, vagal nuclear complex can serve as an explanation for the predominance of the parasympathetic tone of the ANS in patients with a tumor of supratentorial localization.

A few studies were devoted to the analysis of the vegetative status of neurosurgical patients in the perioperative period [41, 44, 49–51]. It was found that the localization of the brain tumor had a significant effect on the vegetative status [51]. Thus, with supratentorial localization of the tumor in the temporal lobe, patients had sympathicotonia with an average level of the personal and high level of situational anxiety. This can be explained both by the direct involvement in the pathological tumor process of the structures of the mediobasal parts of the temporal lobes (amygdala, hippocampus) according to the neuroimaging data presented in the medical history and by indirect irritation of the brain structures forming the limbic system of the brain. In patients with frontal lobe tumors, there was a predominance of sympathicotonia with a high level of personal and an average level of situational anxiety on the eve of surgery. It is known that central noradrenergic systems (in particular, the structures of the brainstem—locus coeruleus) play a significant role in the occurrence of vegetative disorders with pronounced anxiety and fear. Through the ascending pathways, this zone has a connection with both the hypothalamic-pituitary system and the structures of the limbic-reticular complex (hippocampus, amygdala, frontal cortex). Through the descending pathways, noradrenergic structures are connected to the peripheral parts of the sympathetic nervous system. Irritation of the frontal lobes due to tumor growth probably explains the activation of the sympathetic link of the ANS in this category of patients.

3.3 Assessment of the vegetative and psycho-emotional status of neurosurgical patients before operation

The inclusion in the preoperative examination of an anesthesiologist of methods of functional and dynamic examination of the autonomic nervous system to determine the tone of the sympathetic and parasympathetic links of the autonomic nervous system before surgical treatment and assessment of psycho-emotional status [31, 51–54] in elective neurosurgical patients, pain syndrome assessment with the help of VAS of pain allows the anesthesiologist to prescribe an individual premedication to create a vegetative-stabilizing effect, anxiolysis, reducing the afferent flow of information to the brain to create a functional rest of the central nervous system before surgery.

The result of effective premedication will be a smooth induction of anesthesia and a satisfactory intraoperative state of the brain.

DOI: http://dx.doi.org/10.5772/intechopen.101829

Combination option	Premedication evening before surgery	Premedication morning on the day of surgery 30 minutes before admission to the operating room
Eutonia + normal anxiety level	Non-benzodiazepine anxiolytic (hydroxyzine) 25 mg orally	Benzodiazepine anxiolytic phenazepam (bromdihydrochlorphenylbenzodiazepine) 1 mg (1.0 ml 0.1% solution) i/m + NSAIDs (ketoprofen) 100 mg i/m
Parasympathicotonia + average anxiety level	Benzodiazepine anxiolytic phenazepam (bromdihydrochlorphenylbenzodiazepine) 1 mg (1.0 ml 0.1% solution) i/m	Benzodiazepine anxiolytic phenazepam (bromdihydrochlorphenylbenzodiazepine) 1 mg (1.0 ml 0.1% solution) i/m + NSAIDs (ketoprofen) 100 mg i/m
Sympathicotonia + high anxiety level	Benzodiazepine anxiolytic phenazepam (bromdihydrochlorphenylbenzodiazepine) 1 mg (1.0 ml 0.1% solution) i/m + I_1-imidazoline receptor agonist (moxonidine) 0.2 mg orally	Benzodiazepine anxiolytic phenazepam (bromdihydrochlorphenylbenzodiazepine) 1 mg (1.0 ml 0.1% solution) i/m + I_1-imidazoline receptor agonist (moxonidine) 0.2 mg orally + NSAIDs (ketoprofen) 100 mg i/m

Table 1.
Premedication schemes for elective neurosurgical patients depending on the level of personal anxiety and the initial tone of the VNS links.

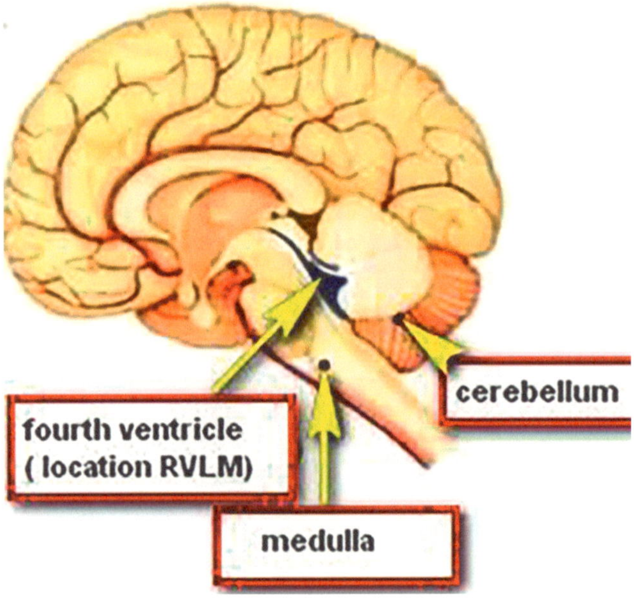

Figure 4.
RVLM and central adrenergic and imidazoline 1-receptors location [55].

Recommended premedication schemes for elective neurosurgical patients, depending on the level of personal anxiety and the initial tone of the ANS links, are presented in **Table 1** [51].

According to modern concepts, the therapy of hyperactivity of the sympathetic nervous system is carried out by influencing the centers that control the work of the cardiovascular system and are located in the brain stem, the most important of which, apparently, is the rostral-ventrolateral region of the medulla oblongata (RVLM) [55]. Various types of receptors are located in this zone, including α2-adrenergic receptors and imidazoline receptors (**Figure 4**) [55]. It has been shown that imidazoline receptors of subtype 1 (I_1) located in RVLM take an active part in blood pressure control, exerting a significant regulatory effect on the activity of the sympathetic nervous system [55].

4. General anesthesia and ANS: control and management in neurosurgery

4.1 Anesthesia depth control (BIS monitoring)

The key anatomic structures of the central nervous system (CNS) that contribute to the state of consciousness are—the brain stem, the pons, the thalamus (thalamic nuclei), and the brain cortex with their connecting neural pathways [56].

General anesthetics (propofol, sevoflurane, desflurane) inhibit the excitatory arousal pathways originating in the brain stem and pons or potentiate the sleep pathways that control them [57]. The brain stem and pontine nuclei have been known to be essential in maintaining cortical arousal and forming the so-called ascending reticular formation [57].

The most important task of an anesthesiologist during neurosurgical operation is to keep anesthesia depth sufficient for security vegetative stability of the patient

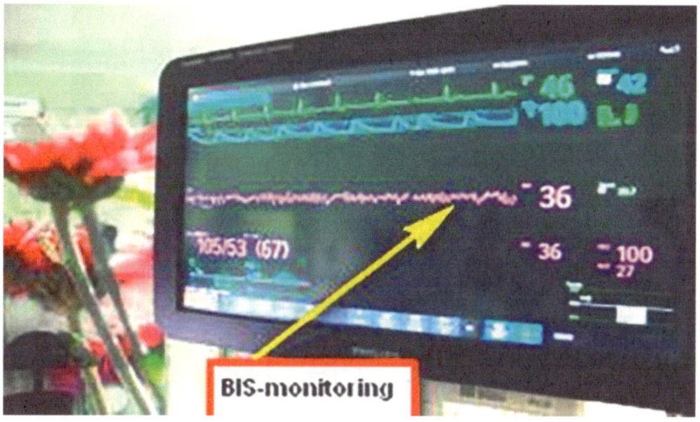

Figure 5.
BIS-monitoring (rose line, BIS = 36) during general anesthesia (TIVA) in patient with brain tumor removal (the image is taken from the private archive of I. A. Savvina).

under the condition of general anesthesia during neurosurgical manipulations especially on the brain stem anatomic structures, cerebral arteries (clipping of cerebral aneurysm), etc. For this purpose, it is necessary to use BIS monitoring. The main component of the BIS monitor is the bispectral analysis, which evaluates the phase relations from a single-channel EEG signal measured from the patient's forehead. The BIS index is a dimensionless number from 0 to 100. For neurosurgery, the optimal means of BIS index is around 40 so during total intravenous anesthesia (TIVA with propofol, opioid μ-agonist fentanyl) as inhalational general anesthesia (sevoflurane, desflurane). Control of the depth of TIVA during brain tumor removal (giant trigeminal schwannoma) is reflected in **Figure 5**.

4.2 Intraoperative antinociceptive defense

Nociceptive stimulation during neurosurgical interventions on the brain, spine and spinal cord, peripheral nerves triggers activation of the sympathetic link of the autonomic nervous system, aseptic systemic inflammatory response [29, 34, 58]. As you know, the actual "pain" receptors are located in the skin, periosteum, and dura mater. Anticipating the development of such a scenario is possible when drugs from the NSAID group are included in the premedication scheme, in particular, ketoprofen, ketonal, lornoxicam [59]; the use of the technique of locoregional anesthesia in neurosurgery [60, 61]. There are both proponents of this approach and its opponents who associate the administration of NSAIDs in the perioperative period in neurosurgical patients with their negative effect on platelet aggregation and additional risks of postoperative hemorrhagic complications [59].

The concept of multimodal multicomponent analgesia in neurosurgical practice finds successful implementation in the form of proactive administration of NSAIDs at the stage of premedication (before the skin incision), locoregional anesthesia using naropin, at the stage of induction and maintenance of anesthesia—opioid analgesic (fentanyl) and α2-adrenergic agonist (clonidine or dexmedetomidine), when suturing a skin wound—paracetamol [53, 62–64].

On the main stage of the neurosurgical operation (brain tumor removing, cerebral aneurysm clipping, spinal hord tumor, AVM removing, etc.) antinociceptive defense is carried out with opioid analgesic μ-agonist fentanyl 3.5–5.0 mcg/kg/h + α2-adrenergic agonist clonidine 0.5 mcg/kg/h or dexmedetomidine 0.2–0.4 mcg/kg/h.

Neurovegetative stabilization with such a method and doses of drugs will be sufficient for cupping of central hemodynamic reactions during neurosurgery [65–69].

4.3 Hemodynamic monitoring

Hemodynamic monitoring is necessary and mandatory part of intraoperative monitoring according to Helsinki Declaration on Patient Safety in Anesthesiology (2010). Noninvasive, invasive hemodynamic monitoring, on the base of technology "PiCCO,"—all options are used in neurosurgery—arterial blood pressure systolic, diastolic, mean (BP), cardiac index (CI), stroke index (SI), global end-diastolic volume index (GEDVI), stroke volume variability (SVV), left ventricular contractility index (LVCI), total peripheral vascular resistance index (TPVRI), intrathoracic blood volume index (ITBVI), extravascular lung water index (EVLWI), pulmonary vascular permeability index (PVPI) [67, 69–71]. The choice of kind and volume of hemodynamic monitoring in neurosurgery is determined by the patient's condition (ASA classification), operation position, localization of brain tumor, supposed blood loss, etc. The level of neurovegetative stabilization is controlled on the basis of the evaluation of the hemodynamic profile of the patient and depth of anesthesia (**Figure 6**) [71].

4.4 Monitoring of neuromuscular conduction

Monitoring of neuromuscular conduction is mandatory for neurosurgical patients for reliable and deep relaxation because the surgical manipulations on the brain structures and cerebral vessels require absolute immobility and synchronization with apparatus of artificial lung ventilation for warning of the rise of intrathoracic pressure when the scull is still closed (dura mater does not open), especially in the patients with intracranial hypertension. Remember that potentiation of neuromuscular block is possible under the condition of water-electrolytic disorders, acid-base state violations, neuromuscular diseases, hypothermia. Control of residual neuromuscular

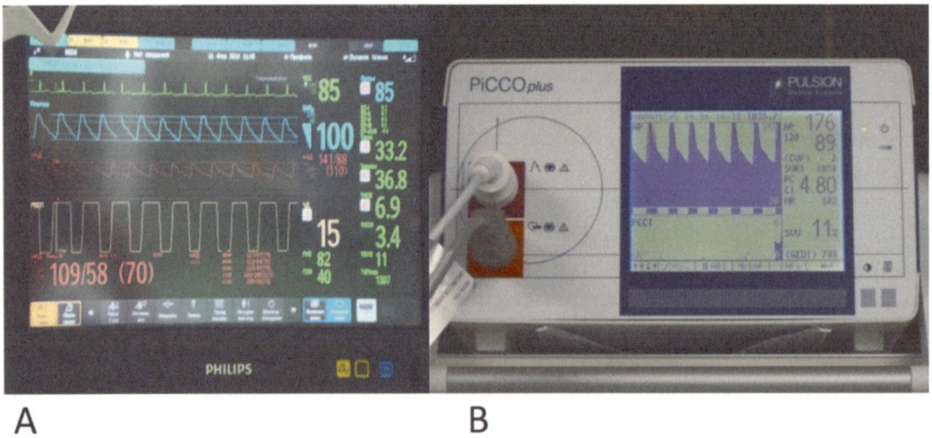

Figure 6.
Intraoperative hemodynamic monitoring during craniofacial block-resection (the image is taken from the private archive of I. A. Savvina). Non-invasive and invasive hemodynamic monitoring on the base of technology "PiCCO": monitor PHILIPS "IntelliVue MX 800" (A); monitor PULSION Medical Systems "PiCCOplus" (B).

block is necessary for decision-making about the extubation of neurosurgical patients in the early postoperative period or the operating room. Musculus adductor pollicis and nervus ulnaris are the most often used for acceleromyography (registration of single twitch (ST); train of four (TOF); post-tetanic count (PTC)).

4.5 Intraoperative thermometry

Intraoperative control of central (rectal, esophageal) and peripheral (skin) temperature is necessary for timely exposure of malignant hyperthermia during inhalational general anesthesia (sevoflurane, for example) and monitoring of balance between heat products and heat dissipation in the patient under the condition of general anesthesia with or without controlled hypothermia. Also, it is necessary when the neurosurgical operation is carried out on the anatomic structures of the third ventricle of the brain and the hypothalamic zone because neurosurgery may cause immediate water-electrolyte disorders and violations of thermoregulation [42].

4.6 Variational cardiointervalometry (heart rate variability)

Variational cardiointervalometry is the noninvasive method of evaluation of the functional state of the cardiovascular system and general condition of the patients and healthy persons. The condition of the vegetative nervous system and mechanisms of regulation of heartbeat is estimated by some statistic, geometric, and special spectral characteristics including R-R intervals, HR, level HR, Baevsky tension index, the balance of the sympathetic and parasympathetic influences (LF/HF), index of centralization (IC), etc. [71, 72]. In neurosurgical patients when calculating the Kerdo index after anesthesia induction and after extubation of patients, a distinct tendency to the state of hypertension was revealed [41]. A significant decrease in heart rate variability was observed in patients with the voluminous formation in the posterior fossa with intracranial hypertension syndrome [50]. It should be noted that the perioperative results of monitoring heart rate variability were obtained using total intravenous anesthesia with the inclusion of α2-adrenergic agonist clonidine, along with dexmedetomidine, used as a component of neurovegetative stabilization in the structure of general anesthesia for neurosurgical interventions on the central nervous system [60, 65, 66, 70, 73, 74]. This method gives the possibility to estimate the prevalence of vegetative tone—sympathetic or parasympathetic or eutonia in neurosurgical patients under the condition of general anesthesia including patients with intracranial hypertension [47–50].

4.7 Method of photoplethysmographic evaluation of the perfusion index

The value of perfusion index PI (N 4–5%) characterizes the volumetric peripheral arterial capillary blood flow [75]. An increase in PI is regarded as excessive perfusion as a result of redistribution of peripheral blood flow and arterioplegia. A decrease in PI values is an early and sensitive sign of adrenergic activity and peripheral vasoconstriction. This indicator is not so informative when assessing the adequacy of anesthesia during the neurosurgical intervention.

The criteria for the adequacy of anesthesiological maintenance in neurosurgery are a volume-stable, moist, pulsating, nonhyperemic brain [67, 70].

Complications	Sitting position	Pron-position	Position on the side	Lounge position
Pulmonary				
Ventilation and perfusion complications	+	+	+	+
Increased pressure in the upper respiratory tract	0	++	0+	0
Tense pneumocephalus	+	+	0	0
Cardiovascular				
Hypotension	++	++	0	+
Rhythm disturbances	++	++	+	++
The need for hemotransfusion	+	++	+−	+
Other complications				
Compression of the eyeballs	0	+++	++	+
Compartment syndrome	+	0	0	0
Venous air embolism	+++	++	+	++
Paradoxical air embolism	++	+	?	?

Table 2.
Complications associated with the position during neurosurgical interventions on the posterior cranial fossa [67].

4.8 Postural circulatory reactions under general anesthesia in neurosurgery

Postural circulatory reactions under general anesthesia in neurosurgery are the most expressed when it is necessary to change horizontal position on the operating table to the operating position "sitting," pron-position, position on the side, lounge position.

Complications associated with the surgical position during neurosurgical interventions on the posterior cranial fossa [67] are presented in **Table 2**.

It is important to estimate the volemic status of the patient, and if the hypovolemia is obtained to start its correction with intravenous infusion of crystalloids (15 ml/kg) and colloid (5 ml/kg) solutions to avoid hypotension during seating of the patient.

4.9 Trigeminocardiac reflex in neurosurgery

The central subtype of trigeminocardiac reflex arises during intracranial impact on the nerve root, central portion of the trigeminal nerve, gasser knot when deep activation of cardiac vagus branch, and depression of lower cardiac sympathetic nerve are discovered. Usually, it is manifested by bradycardia and arterial hypotension [68, 69, 76, 77]. **Figure 7** shows the giant trigeminal schwannoma (the image is taken from the private archive of I. A. Savvina). The removal of this tumor was

DOI: http://dx.doi.org/10.5772/intechopen.101829

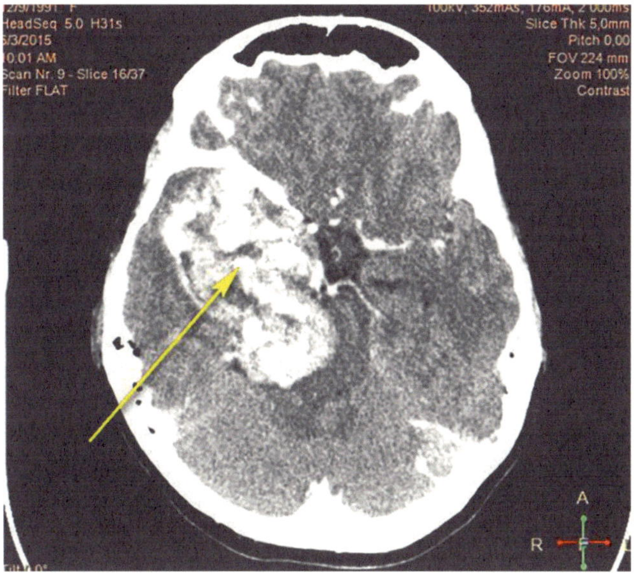

Figure 7.
Giant trigeminal schwannoma (shown by yellow arrow) on CT scan (the image is taken from the private archive of I. A. Savvina).

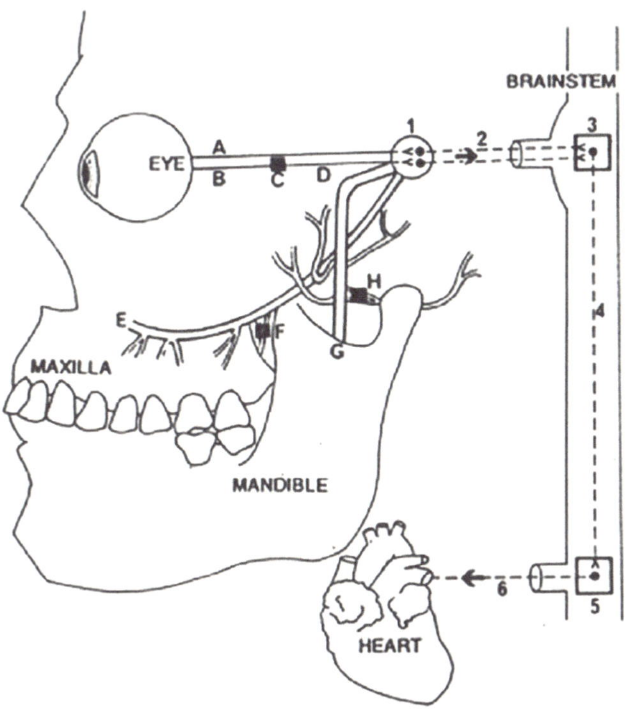

Figure 8.
Pathways of trigeminocardiac reflex [69]: (A) long ciliary nerve; (B) short ciliary nerve; (C) ciliary ganglion; (D) optic nerve; (E) maxillary nerve; (F) winged ganglion; (G) mandibular nerve; (H) auditory ganglion; (1) gasser knot; (2) trigeminal nerve; (3) sensor nucleus of trigeminal nerve; (4) short internuclear fibers; (5) motor nucleus of vagus; and (6) vagus.

accompanied by the central subtype of trigeminocardiac reflex (bradycardia and arterial hypertension as the variant of central TCR) [69, 77].

Figure 8 shows the pathways of trigeminocardiac reflex [69].

5. Vegetative distress-syndrome and pathomorphological signs of insufficiency of ANS

The ANS performs an organizing and trophic function along with endocrine and other anatomical and physiological functional systems [78, 79]. Numerous experimental studies on the plasticity of the nervous system in various injuries of the central nervous system demonstrate the lack of specificity of changes in neurons—in the structure of the neuron nucleus appears invagination of the nuclear membrane, chromatin condensation, swelling of mitochondria, and all cisternal structures [80, 81]. These structural changes appear at any impact on the body, they are a universal ultrastructural expression of the general adaptation syndrome [78]. It is known that the structural and functional plasticity of the nerve cells is unusually high, but the appearance of morphological alterations occurs earlier and persists much longer than functional changes. Pathomorphological studies of the ANS performed on autopsy material of deceased neurosurgical patients revealed severe dystrophic and destructive changes at all levels of the ANS [82].

5.1 Afferent department of visceral reflexes

A pathomorphological study of the structures of the ANS afferent department (receptor apparatuses and sensitive nerve fibers (dendrites) that perceive and conduct afferent impulses), spinal (Th2–Th4; L1–L4; S2–S3), and similar cranial nerve ganglia (trigeminal, inferior vagus node) revealed that regardless of the nature of acute cerebral damage, there are widespread and irreversible violations of the structure and function of the components of the afferent department [28, 82].

Thus, we are talking about the partial or complete death of the sensitive nervous apparatus, the state of which largely determines the reactivity and plasticity in the implementation of an adaptive reaction. Similar changes are found not only in the intramural plexuses but also in the trunk of the vagus nerve, which carries afferent and preganglionic fibers to intramural neurons; in the posterior roots of the spinal cord, where the axons of the neurons of the sensitive spinal ganglia pass. These are nonspecific reactive changes (marginal chromatolysis with preservation of the nucleus, central chromatolysis with preservation of the nucleus, central chromatolysis with "sintering" of lumps of Nissl substance along the periphery of neurons, hyperchromatosis of the nuclei and cytoplasm of the cell in combination with edema and without it), as well as destructive irreversible phenomena (karyolysis with wrinkled hyperchromic nucleus, hydropic changes with the formation of vacuoles and karyolysis, karyorexis in combination with swelling of the neuron body). A motley kaleidoscope of changes reflects the stages of neuronal death depending on local conditions (for example, the acidity of the environment, the degree of hypoxia, hydration of the ganglion, etc.).

5.2 Central parts of the autonomic nervous system

The central parts of the ANS are associative (insertion) links of visceral reflexes. The associative link of the sympathetic nervous system is represented by the nuclei of the lateral horns of the gray matter of the spinal cord in the thoracolumbar region. During spinal cord examination at the level of Th2–Th4, Th12, and L1–L2; the insertion link of the parasympathetic nervous system at the mesencephalic (Yakubovich-Westphal-Edinger nucleus), bulbar (vegetative vagus nerve nucleus), and sacral (S2–S3) levels in the associative links, regardless of the etiological factor, widespread damage to structures and, presumably, disorders of the function of nerve cells, which are mostly irreversible, were noted [74]. Thus, we are talking about partial or significant damage to the associative link, from which the efferent vegetative pathway begins.

Neurites (axons of associative neurons) of the peripheral nerves reach the autonomic ganglia, where they end with synapses. Thanks to synapses, all the links of the visceral reflexes are interconnected and, if necessary, can act as a whole. Significant reactive changes were also revealed in the synaptic apparatus of the associative links (sympathetic and parasympathetic)—argyrophilia (affinity of synaptic rings to nitric acid silver) and hypertrophy of synapses. Destructive changes in the form of fragmentation, granular-lumpy decay of presynaptic nerve fibers, and synaptic structures formed by them on associative neurons are more pronounced in long-term critical conditions of the patient, clinical manifestations of sympathetic hyperactivity syndrome [28, 82, 83]. Nonspecific reactive changes (central and peripheral chromatolysis, acute cellular swelling, process staining, hyperchromatosis of nuclei and cytoplasm) were detected in the central parts of the sympathetic and parasympathetic nervous system. Irreversible destructive phenomena were standard (karyocytolysis, karyolysis). They reflect the different stages of neuronal death. There was also a lively glial reaction [83].

Thus, from a morphological point of view, the most dramatic situation develops in the associative link of the sympathetic nervous system.

The death of neurons nuclei in all parts of the hypothalamus (large-cell nucleus of the anterior hypothalamus, small-cell nuclei of the middle hypothalamus, nuclei, and pathways of the posterior) was also noted [83].

5.3 Efferent section of the visceral reflexes

The efferent autonomic pathway is represented by neurites of associative neurons (preganglionic fibers) and effector neurons and their neurites (postganglionic fibers). The latter reach the innervated tissues, where they realize their impact. The studied sympathetic ganglia (cervical-thoracic or stellate, 2nd–6th thoracic paravertebral, abdominal plexus) are connected by preganglionic fibers with sympathetic centers of the spinal cord; parasympathetic ciliary node—with the Yakubovich-Westphal-Edinger nucleus; Auerbach and Meissner plexuses—with the autonomic nucleus of the vagus nerve.

In acute cerebral injury, extensive structural and functional disorders of the main components of the autonomic ganglion—neurocytes are revealed; at the same time, the “management” of special functions of some organs (glands, smooth muscles of internal organs and vessels, heart muscle, ciliary and pupillary muscles, etc.) and the general adaptive and trophic function suffer [83]. A motley pattern of reactive changes is observed in the sympathetic ganglia—central and peripheral chromatolysis, total chromatolysis with preservation of the nucleus and nucleolus and “sintering” of chromatophilic substance along the edge of the neuron body, hyperchromatosis of the nucleus and perinuclear edema, wrinkling of nuclear material into a homogeneous structureless mass with total chromatolysis (**Figure 9**).

Reactive changes in the parasympathetic ganglia were nonspecific—destructive changes were standard (karyolysis, karyocytolysis), a pronounced reaction from the

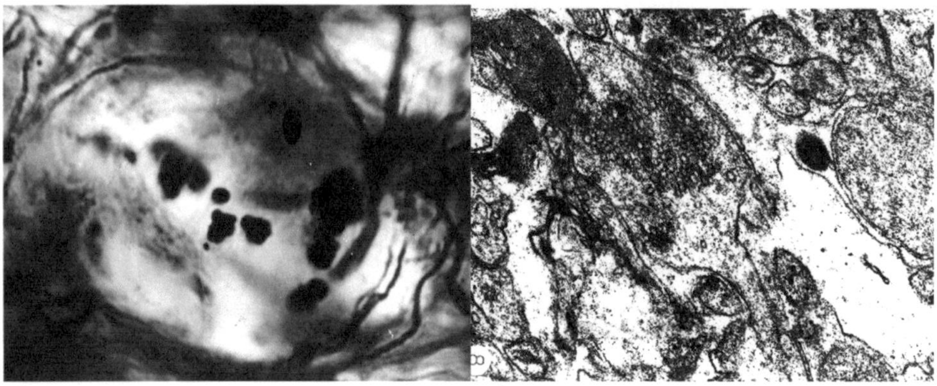

Figure 9.
Sympathetic ganglion 3 days after severe traumatic brain injury (column intermediolateral medullae spinalis). (A) Argyrophilia, hypertrophy, deformation of synapses on neurons. Impregnation by Kahal. Magnification X 100. (B) Destructive changes in the synaptic vesicle. Electronogram. Magnification X 1800.

glia was noted in the ganglia (the number of glial cells (satellites) closely adjacent to the perikaryon of the neurocyte and penetrating it increased around the neurocytes).

The last link of the visceral reflexes is postganglionic fibers, which, as part of nerve trunks and bundles, penetrate all organs and tissues of the body, where they form effector nerve endings. The efferent section of the visceral reflex arches also detects gross dystrophic changes.

The duration of the patient's stay in critical condition correlates with the number of damaged and dead neurons—the longer the period, the more widespread the damage to the ANS was. The death of neurons and their processes can also be caused by their functional overload, excessive functional stress, malfunctions in the rhythm of work, etc.

Some patients who have suffered severe critical conditions recover and return to life with a deeply disabled ANS with a sharply narrowed range of adaptive reactions [28].

Death can occur from the disintegration of the organism as a system with the relative safety of the components of the system (organs) with far from exhausted reserves [83].

Thus, the totality of dystrophic and necrobiotic changes detected in the ANS is the morphological equivalent of vegetative distress syndrome and ANS insufficiency.

6. Conclusion

To summarize, it is obvious that the autonomic nervous system is one of the main systems of life support. Control and monitoring of its functional activity is especially important when the patient is under the condition of general anesthesia. Neurosurgery causes specific central hemodynamic reactions related to reflexes from the brainstem. Management of autonomic reactions and vegetative tone are possible with neurovegetative stabilization and control of the depth of anesthesia.

Conflict of interest

The authors declare no conflict of interest.

Notes/thanks/other declarations

We would like to thank our parents who supported us when we chose medicine as profession and taught us mercy and compassion.

Abbreviations

ANS	autonomic nervous system
CNS	central nervous system
NSAIDs	nonsteroid anti-inflammatory drugs
PONV	postoperative nausea and vomiting
POD	postoperative delirium
PI	perfusion index
TCR	trigeminocardiac reflex
VAS	visual-analog scale of pain

Author details

Irina Alexandrovna Savvina*, Anna Olegovna Petrova and Yulia Mikhailovna Zabrodskaya
Almazov National Medical Research Centre of Ministry of Health of Russian Federation, Polenov Russian Scientific Research Institute of Neurosurgery – Branch of Almazov National Medical Research Centre of Ministry of Health of Russian Federation, Saint Petersburg, Russia

*Address all correspondence to: irinasavvina@mail.ru

IntechOpen

© 2022 The Author(s). Licensee IntechOpen. This chapter is distributed under the terms of the Creative Commons Attribution License (http://creativecommons.org/licenses/by/3.0), which permits unrestricted use, distribution, and reproduction in any medium, provided the original work is properly cited.

References

[1] Becke K. Anesthesia for ORL surgery in children. GMS Current Topics in Otorhinolaryngology - Head and Neck Surgery. 2014;**13**:1-16

[2] Quintao VC. Intraoperative clonidine to prevent delirium following sevoflurane in pediatric patients. Brazilian Journal of Anesthesiology. 2021;**71**(1):4

[3] Nilsson S. Comparative anatomy of the autonomous nervous system. Autonomic Neuroscience. 2011;**16**(165):3-9

[4] Ohtaki H, Shioda S. Anatomy of adult central nervous system: Structure and function of the brain and spinal cord. In: Uchino H, et al. editors. Neuroanesthesia and Cerebrospinal Protection. Japan: Springer; 2015. pp. 3-22. DOI:10.1007/978=4=431-54490-6_1

[5] Benarroch EE. Physiology and pathophysiology of the autonomous nervous system. CONTINUUM: Lifelong Learning in Neurology. 2020;**26**(1):12-24. DOI: 10.1212/CON.0000000000000817

[6] Gibbons CH. Basics of autonomous nervous system function. Handbook of Clinical Neurology. 2019;**160**:407-418. DOI: 10.1016/B978-0-444-64032-1.00027-8

[7] Wehrwein EA. Overview of the anatomy, physiology, and pharmacology of the autonomous nervous system. Comprehensive Physiology. 2016;**13**(6): 1239-1278

[8] Yamashita T, Miyazaki K, Abe K. Cerebrospinal blood flow and its regulation. In: Uchino H et al., editors. Neuroanesthesia and Cerebrospinal Protection. Japan: Springer; 2015. pp. 25-31

[9] Babiyants AY. Cerebral circulation: Physiological aspects and modern research methods. Journal of Fundamental Medicine and Biology. 2018;**3**:46-54

[10] Goadsby PJ. Autonomous nervous system control of the cerebral circulation. Handbook of Clinical Neurology. 2013;**117**:193-201

[11] Sgoifo A, Carnevali L, Alfonso Mde L, et al. Mario amore autonomous dysfunction and heart rate variability in depression. Stress. 2015;**18**(3):343-352. DOI: 10.3109/10253890.2015.104586

[12] Mineur YS, Obayemi A, Wigestrand MB, et al. Cholinergic signaling in the hippocampus regulates social stress resilience and anxiety- and depression-like behavior. Proceedings of the National Academy of Sciences of the United States of America. 2013;**110**(9):3573-3578

[13] Shuleshova NV. Brain Stem: Clinical and Pathophysiological Correspondences. 2nd ed. St. Petersburg: Folio; 2016. p. 356

[14] Reutov VP. New ideas about the role of the autonomic nervous system and nitric oxide generation systems in brain vessels. Pacific Medical Journal. 2016;**2**:10-20

[15] Koolhaas JM, Bartolomucci A, Buwalda B, et al. Stress revisited: A critical evaluation of the stress concept. Neuroscience and Biobehavioral Reviews. 2011;**35**(5):1291-1301

[16] Delefosse MS. Le stress, maladie du siècle? Centre Permanent pour la Citoyenneté et la Participation (CPCP). Au Quotidien. 2017;**1**(19):3-18

[17] Qian X, Droste SK, Gutièrrez-Mecinas M, et al. Rapid release of

corticosteroid-binding globulin from the liver restrains the glucocorticoid hormone response to acute stress. Endocrinology. 2011;**152**(10):3738-3748

[18] Belujon P. Hippocampus, amygdala, and stress: Interacting systems that affect susceptibility to addiction. Annals of the New York Academy of Sciences. 2011;**1216**(1):114-121

[19] Dolzhikov AA, Bobyntsev II, Belykh AE, Dolzhikova IN. Stress, corticosteroid damage to the hippocampus and neuropsychiatric pathology. Kursk Scientific and Practical Bulletin "Man and His Health". 2017;**2**:98-105

[20] Fanselow MS. Are the dorsal and ventral hippocampus functionally distinct structures? Journal of Neuron. 2010;**65**(1):7-19

[21] Colombo J, Arora R, DePace NL, Vinik AI. Clinical Autonomic Dysfunction. Cham: Springer; 2015. p. 452. DOI:10.1007/978-3-319-07371-2

[22] Pershina KV. Neurophysiological mechanisms of stress and depressive conditions and methods of dealing with them. European Science. 2019;**1**(43):78-83

[23] Kozlov AI. Cortisol as a marker of stress. Fiziol People. 2014;**40**(2): 123-136

[24] Gusakov EA. The importance of glucocorticoids in the organization of the stress response of the body. Bulletin of the Vitebsk State Medical University. 2020;**19**(1):24-35

[25] Ueyama T, Senba E. Limbic system and the autonomic nervous system: Electronic record. Rinshō Shinkeigaku. 2010;**50**(11):1003-1006

[26] Selye H. Stress without Distress: "Progress". Moscow. 1982. p. 122

[27] Danilova NN, Krylov AL. Physiology of Higher Nervous Activity. Phoenix: MGU Textbooks; 2002. p. 480

[28] Medvedev YA, Reshetnikov SS, Kondratiev AN. Diagnosis of Vegetative Distress Syndrome on Autopsy Material. St. Petersburg: Russian Polenov Scientific Research Neurosurgical Institute; 1997. p. 21

[29] Bantel C. The role of the autonomous nervous system in acute surgical pain processing—What do we know? Anaesthesia. 2011;**66**(7):541-544

[30] Desborough JP. The stress response to trauma and surgery. British Journal of Anaesthesia. 2000;**85**(1):109-117

[31] Carroll JK, Cullinan E, Clarke L, Davis NF. The role of anxiolytic premedication in reducing preoperative anxiety. British Journal of Nursing. 2012;**21**(8):479-483

[32] Shlyakhto EV, Konradi AO. Causes and consequences of activation of the sympathetic nervous system in hypertension. Arterial Hypertension. 2003;**9**(3):81-88

[33] Golub IE, Sorokina LV. Surgical Stress and Anesthesia. Irkutsk: IGMU; 2005. p. 201

[34] Cortelli P, Giannini G, Favoni V, Cevoli S, Pierangeli G. Nociception and autonomous nervous system. Neurological Sciences. 2013;**34**(1):41-46

[35] Joshi GP. Guidelines for perioperative pain management: Need for re-evaluation. British Journal of Anaesthesia. 2017;**119**(4):703-706

[36] Deryugina AV. Molecular and cellular mechanisms for the implementation

of the stress response of the organism. Proceedings of the UFA Scientific Center of the Russian Academy of Sciences. 2015;**3**:58-63

[37] Gusakova EA. The importance of glucocorticoids in the organization of stress reactions of the body. Bulletin of the Vitebsk State Medical University. 2020;**19**(1):24-35

[38] Gelfand BR. Anesthesiology and Intensive Care. Moscow: Litterra; 2006. p. 576

[39] Mizuno-Matsumoto Y, Inoguchi Y, Carpels SMA, et al. Cerebral cortex and autonomous nervous system responses during emotional memory processing. PLoS One. 2020;**5**(15):3. DOI: 10.1371/journal.pone.0229890

[40] Markin SM. Vegetative dysregulation of blood circulation in patients in the preoperative period: Abstract [dissertation]. St. Petersburg: Sergey Mikhailovich Markin; 2010. p. 23

[41] Lesina SS, Nazarov RV, Kondratiev AN. Features of the autonomic nervous system in patients with brain tumors. Vestnik Rossijskogo universiteta družby narodov. Seriâ: Medicina. 2013;**2**:78-83

[42] Ugryumov VM. Visceral Pathology in Lesions of the Central Nervous System. Leningrad: Medicine; 1975. p. 303

[43] Vinogradov RA. Dynamics of higher mental functions in patients with stenosing lesions of the internal carotid arteries before and after surgical revascularization of the brain. Emergency Medical Care. 2017;**3**:221-227

[44] Racosta JM. Autonomous function and brain volume. Clinical Autonomic Research. 2016;**26**(6):377-383. DOI: 10.1007/s10286-016-0380-8

[45] Freeman R. Testing the autonomic nervous system. Handbook of Clinical Neurology. 2013;**115**:115-136. DOI: 10.1016/B978-0-444-52902-2.00007-2

[46] Kiryachkov YA. Computer analysis of heart rate variability. New opportunities for an anesthesiologist and doctors of other specialties. Vestn Intens Ter. 2002;**1**:3-8

[47] Gorbachev VI. Changes in autonomic homeostasis in the syndrome of intracranial hypertension. Siberian Medical Journal. 2009;**2**:28-30

[48] Gorbachev VI. The relationship of sedation depth with hemodynamic parameters and heart rate variability. Regional Blood Circulation and Microcirculation. 2017;**16**(3):49-55

[49] Smekalov AS. Analysis of the heart rhythm in assessing the level of anesthesia and the physiology of surgical manipulations in the removal of brain tumors: Abstract [dissertation]. St. Petersburg: Alexander Stefanovich Smekalov; 1998. p. 28

[50] Grechaninov EY. Cardiac arrhythmias and methods of their correction during the removal of tumors of the posterior cranial fossa: Abstract [dissertation]. St. Petersburg: Evgeny Yaroslavovich Grechaninov; 1999. p. 23

[51] Orekhova ES. Personalized approach to the choice of premedication in patients with neurosurgical profile. Anesthesiology and Resuscitation. 2019;**3**:79-89

[52] Nishiyama T. Effects of premedication on heart rate variability at induction of anaesthesia: Comparison between midazolam and hydroxyzine. Turkish Journal of Anaesthesiology & Reanimation. 2018;**46**(3):229-232. DOI: 10.5152/TJAR.2018.87059

[53] Naito H. Preoperative assessment. In: Neuroanesthesia and Cerebrospinal Protection. Japan: Springer; 2015. pp. 239-247

[54] Spielberger CD. State-Trait Anxiety Inventory for Adults. New York: Consulting Psychologists Press, Inc.; 2010. p. 76

[55] Michel MC, Ernsberger P. Keeping an eye on the I site: Imidazoline-preferring receptors. Trends in Pharmacological Sciences. 1992;**13**(10):369-370

[56] Mashour GA. Integrating the science of consciousness and anesthesia. Anesthesia and Analgesia. 2006;**103**:975-982

[57] Franks NP. General anaesthesia: From molecular targets to neuronal pathways of sleep and arousal. Nature Reviews. Neuroscience. 2008;**9**:370-386

[58] Gottschalk A. The perioperative management of pain from intracranial surgery. Neurocritical Care. 2009;**10**(3): 387-402

[59] Umamaheswara Rao GS. To use or not to use: The dilemma of NSAIDs and craniotomy. European Journal of Anaesthesiology. 2009;**26**(8):625-626

[60] Kulikov AS. Anesthesiological aspects of accelerated postoperative recovery in neurosurgery: Abstract [dissertation]. Alexander Sergeevich. Moscow: Kulikov; 2020. p. 47

[61] Hansen MS, Brennum J, Moltke FB, et al. Pain treatment after craniotomy: Where is the (procedure-specific) evidence? A qualitative systematic review. European Journal of Anaesthesiology. 2011;**28**(12):821-829

[62] Klimek M, Ubben JF, Ammann J. Pain in neurosurgically treated patients: A prospective observational study. Journal of Neurosurgery. 2006;**104**(3):350-359

[63] Nemergut EC, Durieux ME, Missaghi NB, Himmelseher S. Pain management after craniotomy. Best Practice & Research. Clinical Anaesthesiology. 2007;**21**(4):557-573

[64] American Society of Anesthesiologists Task Force on Acute Pain Management. Practice guidelines for acute pain management in the periodical setting: An updated report by the American Society of Anesthesiologists Task Force on Acute Pain Management. Anesthesiology. 2012;**116**(2):248-273

[65] Mack PF, Perrine K, Kobylarz E, et al. Dexmedetomidine and neurocognitive testing in awake craniotomy. Journal of Neurosurgical Anesthesiology. 2004;**16**(1):20-25

[66] Farag E, Argalious M, Sessler DI, et al. Use of alpha (2)-agonists in neuroanesthesia: An overview. The Ochsner Journal. 2011;**11**(1):57-69

[67] Cottrell JE. Neuroanaesthesia. 6th ed. New York: Elsevier Inc.; 2017. p. 503

[68] Cornelius JF, Sadr-Eshkevari P, Arasho BD, et al. The trigemino-cardiac reflex in adults: Own experience. Expert Review of Cardiovascular Therapy. 2010;**8**:895-898

[69] Savvina IA, Gulyaev DA, Belov IYu, Rutkovskiy RV, et al. Trigeminocardial reflex in skull base surgery. Modern Problems of Science and Education 2017:4. Available from: https://science-education.ru/ru/article/view?id=26631

[70] Kondratiev AN et al. Neuro-Oncology through the Eyes of Anesthesiologists-Resuscitators. St. Petersburg: Publisher of Russian Polenov Scientific Research Neurosurgical

Institute of Ministry of Health of Russian Federation; 2020. p. 229

[71] Heart rate variability: Standards of measurement, physiological interpretation, and clinical use. Task force of the European society of cardiology and the North American Society of Pacing and Electrophysiology. European Heart Journal. 1996;**17**:354-381

[72] Baevsky RM. Analysis of heart rate variability: History and philosophy, theory and practice. Clinical Informatics and Telemedicine. 2004;**1**:54-64

[73] Huang R, Chen Y, Yu AC, et al. Dexmedetomidine-induced stimulation of glutamine oxidation in astrocytes: A possible mechanism for its neuroprotective activity. Journal of Cerebral Blood Flow and Metabolism. 2000;**20**(6):895-898

[74] Gunter A. Intraoperative neurophysiologic monitoring: Utility and anesthetic implications. Current Opinion in Anaesthesiology. 2016;**29**(5):539-543

[75] van Genderen ME, Bartels SA, Lima A, et al. Peripheral perfusion index as an early predictor for central hypovolemia in awake healthy volunteers. Anesthesia & Analgesia 2013;**116**(2):351-356

[76] Chowdhury T, Sandu N, Meuwly C, Cappellani RB, Schaller B. Trigeminal cardiac reflex: Differential behavior and risk factors around the course of the trigeminal nerve. Future Neurology. 2014;**9**:41-47

[77] Savvina IA, Gulyaev DA, Belov IY, Rutkovskiy RV, Smirnova OP, Vaskova NL, Kim AA. Hemodynamic reactions and it's meaning in craniofacial neurosurgery. International Journal of Applied and Fundamental Investigations 2017;**7**:78-83

[78] Anokhin PK. Principal questions of general theory of functional systems. In: Principles of Systemic Organization of Functions. Moscow: Science; 1973. pp. 5-61

[79] Purkayastha S, Stokes M, Bell KR. Autonomic nervous system dysfunction in mild traumatic brain injury: A review of related pathophysiology and symptoms. Brain Injury. 2019;**33**(9): 1129-1136

[80] Gulyaeva NV. Molecular mechanisms of neuroplasticity: An expanding universe. Biochemistry. 2017;**82**(3):237-242

[81] Griesbach GS, Hovda DA. Cellular and molecular neuronal plasticity. Handbook of Clinical Neurology. 2015;**128**:681-690

[82] Medvedev YA. Tanatological Analysis in Pathologo-anatomical Practice. Saint Petersburg: Russian Polenov Scientific Research Neurosurgical Institute of Ministry of Health of Russia; 2005. p. 176

[83] Medvedev YA, Reynus KB. Hypothalamus in neurosurgical patients (pathomorphological and electrophysiological correlations). Archives of Pathology. 1989;**6**:37-43

Chapter 6

Heart Autonomic Nervous System: Basic Science and Clinical Implications

Elvan Wiyarta and Nayla Karima

Abstract

The heart has an intrinsic conduction system that consists of specialized cells. The heart receives extensive innervation by both sympathetic and parasympathetic systems of the ANS. The ANS influences most heart functions by affecting the SA node, AV node, myocardium, and small and large vessel walls. The sympathetic system carries an excitatory effect on heart functions. Conversely, the parasympathetic system has inhibitory effects on heart functions. ANS abnormalities in terms of anatomy and physiology can cause various heart abnormalities. ANS abnormalities associated with electrical abnormalities can cause a variety of heart manifestations. Besides electrical abnormalities, ANS also correlates with ischemic heart disease. Following electrical and ischemic instability, ANS also have direct effect on action potential duration restitution. By understanding the mechanism of influence of the anatomy and physiology of the ANS heart and its influence on various heart abnormalities, we can determine the appropriate therapeutic approaches. Therapeutic approaches in neurocardiology fall into two focuses: applying novel treatment and interaction of non-drug and multiple drugs treatments.

Keywords: heart, sympathetic, parasympathetic, autonomous, neurology

1. Introduction

The heart has an intrinsic conduction system that consists of specialized cells. It can spontaneously depolarize to initiate heartbeats from its rhythmic pacing discharge and coordinate heart electrical activity [1, 2]. The sinoatrial (SA) node is the first pacemaker that starts the electrical impulse resulting in the depolarization and contraction of the atrium. This electrical impulse is distributed throughout the heart through the internodal pathway, atrioventricular (AV) node, AV bundle, branches of the bundle of HIS, and through Purkinje fibers. Without the extrinsic (hormonal and neural) influences, the SA node creates about 100 beats per minute; however, to meet the body's oxygen requirement under variable conditions, cardiac output (and thus heartbeat) must vary. This is where the autonomic nervous system (ANS) of the heart plays a role [2].

2. Basic science in the ANS of the heart

The heart receives extensive innervation by both sympathetic and parasympathetic systems of the ANS. The cardiac efferent preganglionic sympathetic neurons originate from the lateral horns of the spinal cord's upper thoracic segment (T1-T4) and leave the spinal cord through the ventral (anterior) roots of the corresponding spinal cord nerves. As they reach the superior cervical, medial cervical, cervico-thoracic/stellate, and thoracic ganglia of the paravertebral sympathetic nerve chain (SNC), they synapse onto the postganglionic nerves, namely the cardiac cervical nerves and cardiac thoracic nerves, which travel to the heart along with the epicardial vascular structure [1–4].

The cardiac efferent preganglionic parasympathetic neurons originate in the medulla oblongata's dorsal motor nucleus and nucleus ambiguus. They travel bilaterally within two vagal nerves and synapse onto the postganglionic nerve fibers in the vagal nerve ganglia located in the cardiac plexus, at the base of the heart [3, 4]. Cardiac plexus consists of a complex network of various nerves including the sympathetic, parasympathetic, and cardiac nerves as well as some tiny parasympathetic ganglia to control cardiac activity. The cardiac plexus is divided into two parts: (1) the superficial part located in the aortic arch concavity and (2) the deep part located between the trachea and the aortic arch. Both parts are connected to provide cardiac autonomic innervation [3].

Most of the cardiac afferent fibers travel in sympathetic cardiac nerves. The first-order sympathetic-sensitive afferent fibers have their cell bodies in the first 4–5 thoracic ganglia. They synapse with the second-order fibers in the spinal cord, where they cross the median line and ascend along the anterior spinothalamic tract (ventral spinothalamic fasciculus) to the posteroventral nucleus in the thalamus. Parasympathetic afferent fibers in the heart primarily function as a mediator for some cardiac reflexes, responding to activation of stretch receptors in the atria (Bainbridge reflex) and left ventricle (Jarisch-Bezold reflex) [3].

The ANS influences most heart functions by affecting the SA node, AV node, myocardium, and small and large vessel walls [2]. The ANS regulates heart rate (chronotropic effect), myocardial cells contractility (inotropic effect), signal conductivity (dromotropic effect), excitability (bathmotropic effect), as well as coronary vascular tone and myocardial blood flow. As the sympathetic and parasympathetic systems have opposite effects on heart functions, the final effect on the heart is the net balance between the two systems. However, their influence differs by their distribution in the heart [2, 3].

The sympathetic system carries an excitatory effect on heart functions and is activated in emergency, stressful situations, or any other situations that require increase of cardiac output; therefore, it is also known as "fight or flight response" [2]. It controls heart function mainly in three effects: (1) It speeds up the depolarization of the sinus node increasing heart rate (positive chronotropic), (2) increases conduction velocity in the AV junction, atria, and ventricles (positive dromotropic effect), (3) increases myocardial contractility both in the atria and ventricle (positive inotropic effect) [2, 3]. Most of these effects are mainly mediated by the β1 adrenergic receptors as they predominate in healthy human hearts, whereas β2 receptors are primarily concentrated in the atria and ventricles thus their functions are linked to the inotropic effect. Both β1 and β2 receptors are distributed in all regions of the heart, nevertheless [3]. In addition, sympathetic activation also promotes constriction of the coronary arteries leading to an increase of cardiac output, which is mediated by α1 and α2 receptors, and dilatation mediated by β2 receptors in the coronary arteries [2, 3].

Conversely, the parasympathetic (vagal) system has inhibitory effects on heart functions. It is activated under restful conditions and is therefore known as rest and digest response [2]. It slows down sinus node activity resulting in a decrease of heart rate, slows down electrical conduction through the AV nodes and conduction system, causing delayed conduction and AV block, decreases atria contractility, and promotes dilatation of the coronary arteries, which result in decreased cardiac output. On atrial cells, parasympathetic activation decreases contractility yet shortens the action potential duration causing an increase in conduction speed, thus leading to reentrant tachyarrhythmias. As parasympathetic fibers are predominantly distributed to the atria while poorly distributed to the ventricles, parasympathetic activation does not significantly affect intraventricular conduction and ventricles' contractility. The parasympathetic system influences the heart through the M2 receptor and the coronary arteries through M3 receptors [3].

Both sympathetic and parasympathetic preganglionic neurons release acetylcholine (Ach) and are called cholinergic; however, their postganglionic release different neurotransmitters. Sympathetic postganglionic neurons release norepinephrine (which resembles epinephrine/adrenalin, thus referred to as adrenergic) while most parasympathetic postganglionic neurons release acetylcholine.[3]

3. Influence of ANS on electrical abnormalities in heart

ANS abnormalities in terms of anatomy and physiology can cause various heart abnormalities. ANS abnormalities are associated with electrical abnormalities which cause heart problems. This can cause a variety of manifestations. In this section, we will discuss more the electrical abnormalities associated with ANS abnormalities in the heart.

3.1 Ventricular arrhythmias

Ventricular arrhythmia remains a common cause of sudden cardiac death in myocardial infarction (MI) patients. Following a myocardial ischemic injury, sympathetic axon fibers within the scar become dysfunctional, degenerate, and die. However, contrary to the central neurons, peripheral neurons commonly regenerate back to their target, a phenomenon called nerve sprouting [4, 5]. This efferent sympathetic regeneration is triggered by nerve growth factor (NGF), which levels are found to be increased after MI, and causes hyperinnervation in the infracted are of the heart thereby promoting ventricular arrhythmia. Studies using 123I-meta-iodobenzylguanidine (MIBG) have shown evidence of sympathetic reinnervation in the infracted hearts after MI. A study conducted by Cao et al. [6] demonstrated that the high density of nerve fibers was significantly higher in the peripheral to the area of necrotic tissue of failed hearts. Chen and colleagues also support this phenomenon's discovery that infusion of NGF to the stellate ganglion causes an increase of nerve density and QT interval prolongation, therefore increases and prolongs ventricular arrhythmias [4, 6–8]. Furthermore, there have been findings that demonstrate a notable decrease in parasympathetic tone in patients with comorbidities (such as coronary artery disease, MI, and diabetes) during sleep despite the unopposed sympathetic activity, creating a higher risk of ventricular arrhythmia. Another electrical phenomenon following MI that leads to ventricular arrhythmia is an occurrence of heterogeneous distribution of hyperinnervation of sympathetic nerves, particularly in the border zone (despite the remaining viable myocardial cells), which can lead to impulses and therefore initiate tachyarrhythmia. On another note, interventions that reduce sympathetic nerve activity have been shown to reduce the risk of arrhythmias in MI patients, both in humans and animals [6]. Some therapies that are suggested to reduce the risk of ventricular arrhythmia include cervical sympathectomy and spinal cord stimulation (inhibiting cardiac sympathetic tone while enhancing parasympathetic tone). Future therapies may focus on preventing nerve sprouting by inhibiting nerve growth or attaining regional cardiac denervation by ganglia ablation [4].

3.2 Atrial fibrillation

The influence of ANS on the pathogenesis of atrial fibrillation (AF) had been discovered since 1978 [3]. In the beginning, AF was thought to be a sympathetic-mediated phenomenon; however, studies have shown that sympathetic and parasympathetic systems may contribute to the pathogenesis. Sympathetic-mediated arrhythmia may occur because of β-adrenergic signal pathway activation, which increases Ca^{2+} transient. On the other hand, parasympathetic activation through Ach stimulation on muscarinic receptors (mainly M2 in the heart) causes a shortened duration of action potential (thus increasing conduction speed) in atria, causing arrhythmias [4, 9]. Studies by Scherf et al. suggested that local application of either aconitine or Ach in the heart may lead to rapid focal firing or AF, which could be terminated by removing the focal source of firing [10, 11]. Whether an AF episode is predominately sympathetic-mediated or parasympathetic-mediated may depend on comorbidities; lone and nocturnal AF (where parasympathetic is profoundly dominant) in patients with normal hearts is usually parasympathetic-mediated whereas AF in patients with organic heart disease or disorders such as phaeochromocytoma or hyperthyroidism is usually sympathetic-mediated. In addition, parasympathetic-mediated AF episodes usually occur weekly, predominantly at night, last for a few hours, and are preceded by progressive bradycardia. In contrast, sympathetic-mediated AF episodes usually occur during the daytime, during exercise, or under stress. The current primary endpoint target of the ablation procedure is the pulmonary vein isolation (PVI), thereby predisposing to reentrant phenomena and high density of nerves. However, studies have demonstrated that direct stimulation to the ganglionated plexus could result in AF, whereas ablation of the corresponding plexus may reverse the alteration of conduction speed [3, 8]. Multiple clinical studies were conducted to compare whether combining ganglionated plexus (GP) ablation with PVI or PVI alone is more effective in suppressing AF, one of which is done by Katritsis et al. l who found that combination of GP ablation and PVI showed higher success compared to PVI alone [9].

3.3 Long QT syndrome

Long QT syndrome (LQTS) is characterized by prolonged ventricular repolarization (prolonged QT interval), leading to polymorphic ventricular tachycardia and, therefore, risk of sudden death. It is a heterogeneous syndrome resulting from several cardiac ion channels. Arrhythmias in LQTS patients are often emotional or physical stress-related, and sympathetic activation has been suggested as an important triggering factor. However, the response to this trigger may vary depending on LQTS syndrome. For instance, LQTS type 1 has more prominent and prolonged effects from sympathetic activation than LQTS type 2 [4]. A study has been conducted by

Shamsuzzaman [12] to record sympathetic activity using muscle sympathetic nerve activity (MSNA) and skin sympathetic nerve activity (SNA). The result of the study demonstrated that in LQTS patients, the baseline of MSNA is very low and further accompanied by slower heart rates and reduced LF. In contrast, the baseline of skin SNA is normal, indicating that LQTS patients have region-specific decreased cardiac sympathetic drive. In such a setting, surges of sympathetic stimulation caused by emotional or physical stress may lead to cardiovascular events [12].

3.4 Brugada syndrome

Brugada syndrome is an inherited channel disorder characterized by sodium channel abnormality (and thus ECG abnormalities) that predisposes to ventricular arrhythmias and sudden death despite structurally typical hearts [4, 13, 14]. Another exciting characteristic of Brugada syndrome is that ventricular fibrillation and sudden death mainly occur at rest or during sleep, which is the period of parasympathetic dominance. Furthermore, clinical characteristics and typical ECG changes can be variable over time and are influenced by external factors, such as exercise and pharmacological intervention. Exercise can diminish ECG signs of Brugada syndrome, while on the contrary, drugs that interact with the ANS innervation can unmask or intensify the signs. For this occurrence, studies have suggested that the ANS is involved in the natural history of the syndrome. Prior studies have shown a sympathetic-parasympathetic tone imbalance in patients with Brugada syndrome. A study by Wichter et al. demonstrated a reduced I-MIBG reuptake, either because of a reduced number or function of efferent sympathetic neurons and a reduced transporter capacity for NE reuptake, which indicated a presynaptic adrenergic dysfunction [14]. According to the authors of this study, this reduced sympathetic tone may impact protein phosphorylation and spatial calcium heterogeneity, thus leading to arrhythmias, especially in the downregulation of adrenergic activity or in parasympathetic dominance [14].

4. Influence of ANS in heart failure and myocardial infarction

Besides electrical abnormalities, ANS also correlates with ischemic heart disease. Following a transmural myocardial infarction (MI), sympathetic fibers within the scar become denervated and die. However, denervation also occurs in the non-infarcted sites distal to the infarction early after occlusion, resulting in a neurotransmission disruption, nerve sprouting, and denervation supersensitivity even in the viable myocardium cells. Not all sites are denervated equally, this disruption leads to a heterogeneous change of effective refractory period (ERP). Together with decreased protection from vagal denervation, this leads to ventricular arrhythmias [4].

As with heart failure, myocardial dysfunction caused by cardiac insult activates neurohormonal mechanisms, including activation of the sympathetic system and the renin-angiotensin-aldosterone system (RAAS) axis. Increased activation of the sympathetic system causes an increase in NE delivery to myocardial cells. High local catecholamine level leads to ventricular hypertrophy and increase susceptibility to arrhythmia, which worsens the heart's function and, in turn, further increases sympathetic tone [15]. This activation is initially essential to compensate for the weakened myocardial function; however, in the long term, this activation leads to further deterioration of cardiac function, worsening heart failure, and cardiac decompensation. Besides sympathetic activation, there has been evidence of reduced parasympathetic function, which further worsens heart failure. Heart failure can also cause denervation, creating nerve sprouting and electrical remodeling, leading to ventricular arrhythmia and sudden cardiac death [4, 16].

5. Effect of the ANS on action potential duration restitution

Following electrical and ischemic instability, ANS also have a direct effect on action potential duration restitution. The destabilization of activation wavefronts is associated with the alteration in action potential duration (APD) resulting from the alteration of the previous diastolic interval, called restitution. Steepened APD restitution curve slope has been associated with complex, unstable dynamics, while a decrease of the steepness of the curve by drugs may suppress ventricular arrhythmia [17–19]. A study in porcine models by Taggart et al. has shown that sympathetic stimulation with adrenaline (α – and β-adrenergic agonist) steepens the APD restitution curve [20]. The same effect was confirmed in humans with normal ventricles by a more recent study using isoprenaline (β-adrenergic agonist) and adrenaline, demonstrating that both adrenaline and isoprenaline steepen the APD restitution curve at the minimum range of 40 ms. This evidence suggests a mechanism in which the sympathetic nervous system is contributed to inducing arrhythmia and ventricular fibrillation [16]. Additionally, a study conducted in an isolated rabbit heart model demonstrated that parasympathetic activation exerts a contradictory effect, reducing the steepness of the slope, thereby suppressing ventricular fibrillation [21].

6. Therapeutic approaches involving ANS in the heart

By understanding the mechanism of influence of the anatomy and physiology of the ANS heart and its influence on various heart abnormalities, we can determine the appropriate therapeutic approaches. Therapeutic approaches in neurocardiology fall into two focuses: (1) applying novel treatment and (2) interaction of non-drug and multiple drugs treatments. Patients with cardiomyopathy are suggested to have increased sympathetic innervation and decreased parasympathetic innervation; therefore, interventions aiming to reduce sympathetic tone and thereby increasing parasympathetic tone are beneficial to reduce the susceptibility of ventricular arrhythmia sudden cardiac death. Some options of approaches include the following options [4].

6.1 Selective sympathetic blockade

Multiple studies have shown that in patients with heart failure, pharmacologically inhibition of sympathetic activity may reduce the risk of sudden cardiac death. Current pharmacological therapies include β-blockers (β-receptor antagonist) and angiotensin-converting enzyme inhibitors (ACE-I), which are the mainstay approaches for early hypertension and other cardiovascular disease associated with dysautonomia [22]. Surgical techniques, for instance, sympathectomy, reduce the risk of comorbidities in patients with hypertension and reduce the incidence of ventricular arrhythmia [22].

6.2 Cardiac autonomics modulation therapies

Pharmacological therapies such as β-blockers, ACE-I, angiotensin receptor blockers (ARB), aldosterone antagonists, and statins are proven to decrease the risk of sudden cardiac death in patients with ischemic cardiomyopathy. In addition, these drugs also provide modulations of the ANS by decreasing sympathetic activity and increasing parasympathetic activity. Through baroreflex, Angiotensin II decreases vagal bradycardia. This effect can be reversed with ACEI and ARB by increasing parasympathetic output to the heart. In an experimental study using rat models with ischemic cardiomyopathy, aldosterone antagonist and ACEI showed a decrease of myocardial NE content, demonstrating an antisympathetic effect. Statin therapies show several mechanisms in normalizing sympathetic activity and cardiovascular reflex regulation, such as increased baroreceptor sensitivity for heart rate control, reducing angiotensin II-induced sympathetic responses, decreasing baseline of renal sympathetic activity, and downregulating mRNA and protein expression of Angiotensin II type I receptors as well as NADP oxidase subunits of the heart [4].

6.3 Resynchronization therapy

Biventricular pacing has been suggested to improve hemodynamic status in patients with intraventricular conduction delay and reduced ejection fraction and decreased sympathetic tone in patients with hypertension, thus shifting the autonomic balance of the heart to a less sympathetic more parasympathetic profile [4]. Proper cardiac resynchronization therapy (CRT), in the short term, results in left ventricular systolic function improvement and mitral regurgitation reduction, providing a more optimal ventricular filling. Over a more extended period, CRT promotes left ventricular reverse remodeling, leading to significant functional capacity, survival, and quality of life improvements [23].

6.4 Parasympathetic function mortality and cardiovascular risk

Several measurements that can be used to index parasympathetic function/activity include resting heart rate, heart rate recovery (heart rate decrease following termination of exercise), heart rate variability, and baroreflex sensitivity (the responsiveness of the cardiovascular system to blood pressure changes). Several studies have shown that reduced parasympathetic function is associated with mortality and leads to risk factors for cardiovascular diseases. Those risk factors include biological factors such as hypertension, diabetes, abnormal cholesterol; lifestyle factors such as tobacco use, physical inactivity, and overweight; and non-modifiable factors such as age and family history [4].

6.5 Vagal stimulation

Vagal nerve stimulation (VNS) is a non-pharmacological intervention to normalize autonomic imbalance, directly stimulating the vagus nerve to improve parasympathetic tone and reflex. VNS has been shown to improve left ventricular hemodynamics and increase heart rate variability. VNS also results in better vagal reflex and nitric oxide expression, improvement of the renin-angiotensin system, inflammatory cytokines modulation, reduced heart rate, risk of ventricular arrhythmias, and mortality [24]. A recent multinational, randomized clinical trial called INOVATE-HF (Increase of vagal tone in CHF) demonstrated that VNS significantly resulted in favorable effects on quality of life, NYHA functional class, and 6-min walking distance. However, the ventricular end-systolic volume index was not significantly different [25].

6.6 Renal denervation

Renal efferent signals regulate renin secretion, water and sodium retention, and intrarenal vascular distribution. Efferent signals (as a response to sensory signals from renal) activate sympathetic fibers, inhibit parasympathetic fibers, and cause a release of catecholamines, which in pathology conditions such as myocardial infarction or heart failure, can increase the risk of arrhythmia [26]. Catheter-based renal denervation (RDN) is a neuromodulation treatment that includes catheter-based ablation to the renal artery wall, thus reducing the afferent and efferent sympathetic activity in the kidney and globally [26–28]. It has been used to treat drug-resistant hypertension. However, the role of RDN has also been studied as adjunctive therapy in patients with ventricular tachycardia and heart failure. By reducing circulating catecholamines, RDN reduces the electrical heterogeneity in the scarred myocardium and border zone regions and thus decreases susceptibility to ventricular arrhythmia and sudden cardiac death [26]. RDN has also been suggested to reduce blood pressure, reduce NT-proBNP, and improve NYHA class symptoms in patients with heart failure. Therefore, RDN is suggested to be favorably impactful for hypertension, MI, and heart failure [28].

7. Conclusion

The heart receives extensive innervation by both sympathetic and parasympathetic systems of the ANS. The sympathetic system carries an excitatory effect on heart functions, while the parasympathetic system has inhibitory effects on heart functions. ANS abnormalities associated with electrical abnormalities can cause a variety of heart manifestations, including ventricular arrhythmias, atrial fibrillation, Long QT Syndrome, and Brugada Syndrome. Besides electrical abnormalities, ANS also correlates with ischemic heart disease. Following electrical and ischemic instability, ANS also have a direct effect on action potential duration restitution. By understanding the mechanism of influence of the anatomy and physiology of the ANS heart and its influence on various heart abnormalities, we can determine the appropriate therapeutic approaches. Therapeutic approaches in neurocardiology fall into two focuses: applying novel treatment and interaction of non-drug and multiple drugs treatments, such as selective sympathetic blockade, cardiac autonomics modulation therapies, resynchronization therapy parasympathetic function mortality and cardiovascular risk, vagal stimulation, and renal denervation.

Acknowledgements

No one to acknowledge.

Conflict of interest

The authors declare no conflict of interest.

DOI: http://dx.doi.org/10.5772/intechopen.101718

Appendices and nomenclature

Acetylcholine	neurotransmitters in ANS
ANS	the autonomic nervous system consists of sympathetic and parasympathetic components
Atrial fibrillation	electrical heart abnormalities which are generally characterized by arrhythmias on electrocardiogram findings
AV block	heart abnormalities based on block of the cardiac conduction system in the AV node
AV node	heart node located at the atrioventricular junction
Bainbridge reflex	compensating reaction occurring in an increase in heart rate after an increase in cardiac preload
Baroreflex	compensating reaction occurring in an increase in heart rate after an increase in cardiac preload
Hyperinnervation	excessive innervation
Jarisch-Bezold reflex	bradycardia, hypotension, and apnea
Myocardial infarction	heart abnormalities in the form of damage to heart cells due to lack of blood supply to the cells concerned
Neurocardiology	the branch of neurology that studies the nervous system of the heart
NYHA functional class	The New York Heart Association's (NYHA) functional classification system assists in classifying individuals with congestive heart failure based on their symptoms.
Pulmonary Vein Isolation	a treatment used to treat atrial fibrillation, an irregular heart rhythm.
SA node	a cluster of cells in the right atrium. These cells can deliver electrical impulses to the heart muscle cells, causing them to contract regularly and autonomously.
Sympathetic nerve chain	ganglionated chain from the skull base to the coccyx
Vagal nerve stimulation	refers to any procedure that stimulates the vagus nerve, whether physical or electronic.

Author details

Elvan Wiyarta* and Nayla Karima
Faculty of Medicine, Universitas Indonesia, Jakarta, Indonesia

*Address all correspondence to: elvan.wiyarta@ui.ac.id

IntechOpen

© 2022 The Author(s). Licensee IntechOpen. This chapter is distributed under the terms of the Creative Commons Attribution License (http://creativecommons.org/licenses/by/3.0), which permits unrestricted use, distribution, and reproduction in any medium, provided the original work is properly cited.

References

[1] Lemieux J, Edelman E, Strichartz G, Lilly L. Normal cardiac structure and function. In: Lilly L, editor. Pathophysiology of Heart Disease: A Collaborative Project of Medical Students and Faculty. 6th ed. Philadelphia: Wolters Kluwer; 2016. pp. 7-25

[2] Gordan R, Gwathmey JK, Xie LH. Autonomic and endocrine control of cardiovascular function. World Journal of Cardiology. 2015;**7**(4):204-214

[3] Battipaglia I, Lanza G. The autonomic nervous system of the heart. In: Slart R, Tio R, Elsinga P, Schwaiger M, editors. Autonomic Innervation of the Heart. New York: Springer; 2015. pp. 1-11

[4] Asmundis C, Camp G, Brugada P. Electrophysiology and pathophysiology of the autonomic nervous system of the heart. In: Slart R, Tio R, Elsinga P, Schwaiger M, editors. Autonomic Innervation of the Heart. New York: Springer; 2015. pp. 14-56

[5] Huang WA, Boyle NG, Vaseghi M. Cardiac innervation and the autonomic nervous system in sudden cardiac death. Cardiac Electrophysiology Clinics. 2017;**9**(4):665-679

[6] Li CY, Li YG. Cardiac sympathetic nerve sprouting and susceptibility to ventricular arrhythmias after myocardial infarction. Cardiology Research and Practice. 2015;**2015**:698368

[7] Cao JM, Fishbein MC, Han JB, Lai WW, Lai AC, Wu TJ, et al. Relationship between regional cardiac hyperinnervation and ventricular arrhythmia. Circulation. 2000;**101**(16):1960-1969

[8] Chen PS, Chen LS, Cao JM, Sharifi B, Karagueuzian HS, Fishbein MC. Sympathetic nerve sprouting, electrical remodeling and the mechanisms of sudden cardiac death. Cardiovascular Research. 2001;**50**(2):409-416

[9] Xi Y, Cheng J. Dysfunction of the autonomic nervous system in atrial fibrillation. Journal of Thoracic Disease. 2015;**7**(2):193-198

[10] Scherf D, Morgenbesser LJ, Nightingale EJ, Schaeffeler KT. Further studies on mechanism of auricular fibrillation. Proceedings of the Society for Experimental Biology and Medicine. 1950;**73**(4):650-654

[11] Scherf D. Studies on auricular tachycardia caused by aconitine administration. Proceedings of the Society for Experimental Biology and Medicine. 1947;**64**(2):233-239

[12] Shamsuzzaman ASM, Ackerman MJ, Kara T, Lanfranchi P, Somers VK. Sympathetic nerve activity in the congenital long-QT syndrome. Circulation. 2003;**107**(14):1844-1847

[13] Li KHC, Lee S, Yin C, Liu T, Ngarmukos T, Conte G, et al. Brugada syndrome: A comprehensive review of pathophysiological mechanisms and risk stratification strategies. International Journal of Cardiology Heart and Vasculature. 2020;**26**:100468

[14] Wichter T, Matheja P, Eckardt L, Kies P, Schäfers K, Schulze-Bahr E, et al. Cardiac autonomic dysfunction in Brugada syndrome. Circulation. 2002;**105**(6):702-706

[15] Goldstein DS. Neurocardiology: Therapeutic implications for cardiovascular disease. Cardiovascular Therapeutics. 2012;**30**(2):e89-e106

[16] Kishi T. Heart failure as an autonomic nervous system dysfunction. Journal of Cardiology. 2012;**59**(2):117-122

[17] Taggart P, Sutton P, Chalabi Z, Boyett MR, Simon R, Elliott D, et al. Effect of adrenergic stimulation on action potential duration restitution in humans. Circulation. 2003;**107**(2):285-289

[18] Chialvo DR, Gilmour RF Jr, Jalife J. Low dimensional chaos in cardiac tissue. Nature. 1990;**343**(6259):653-657

[19] Garfinkel A, Kim YH, Voroshilovsky O, Qu Z, Kil JR, Lee MH, et al. Preventing ventricular fibrillation by flattening cardiac restitution. Proceedings of the National Academy of Sciences of the United States of America. 2000;**97**(11):6061-6066

[20] Taggart P, Sutton P, Lab M, Dean J, Harrison F. Interplay between adrenaline and interbeat interval on ventricular repolarisation in intact heart in vivo. Cardiovascular Research. 1990;**24**(11):884-895

[21] Ng GA, Brack KE, Coote JH. Effects of direct sympathetic and vagus nerve stimulation on the physiology of the whole heart--a novel model of isolated Langendorff perfused rabbit heart with intact dual autonomic innervation. Experimental Physiology. 2001;**86**(3):319-329

[22] Bardsley EN, Paterson DJ. Neurocardiac regulation: From cardiac mechanisms to novel therapeutic approaches. The Journal of Physiology. 2020;**598**(14):2957-2976

[23] O'Brien T, Park MS, Youn JC, Chung ES. The past, present and future of cardiac resynchronization therapy. Korean Circulation Journal. 2019;**49**(5): 384-399

[24] Camm AJ, Savelieva I. Vagal nerve stimulation in heart failure. European Heart Journal. 2015;**36**(7):404-406

[25] Gold MR, Van Veldhuisen DJ, Hauptman PJ, Borggrefe M, Kubo SH, Lieberman RA, et al. Vagus nerve stimulation for the treatment of heart failure: The INOVATE-HF trial. Journal of the American College of Cardiology. 2016;**68**(2):149-158

[26] Bradfield JS, Vaseghi M, Shivkumar K. Renal denervation for refractory ventricular arrhythmias. Trends in Cardiovascular Medicine. 2014;**24**(5):206-213

[27] Sharp TE 3rd, Lefer DJ. Renal denervation to treat heart failure. Annual Review of Physiology. 2021;**83**:39-58

[28] Fudim M, Sobotka PA, Piccini JP, Patel MR. Renal denervation for patients with heart failure: Making a full circle. Circulation. Heart Failure. 2021;**14**(3):e008301

Chapter 7

New and Emerging Technologies for Integrative Ambulatory Autonomic Assessment and Intervention as a Catalyst in the Synergy of Remote Geocoded Biosensing, Algorithmic Networked Cloud Computing, Deep Learning, and Regenerative/Biomic Medicine: Further Realization of Multidomain Personalized Health

Robert L. Drury

Abstract

While the important role of the autonomic nervous system (ANS) has been historically underappreciated, recently there has been a rapid proliferation of empirical, methodological and theoretical progress in our more detailed understanding of the ANS. Previous more simplistic models of the role of the ANS using the construct of homeostasis have been enhanced by the use of the construct of allostasis and a wide variety of technological innovations including wearable and implantable biosensors have led to improved understanding of both basic and applied knowledge. This chapter will explore in particular heart rate variability (HRV) as a rich variable which has developed an extensive literature, beginning with predicting all-cause mortality, but now encompassing a wide variety of disease and illness states; cognitive, affective and behavioral processes and performance optimization. A critical analysis of HRV from the perspective of complex adaptive systems and non-linear processes will be included and innovative future uses of HRV will be described.

Keywords: ANS, HRV, Digital epidemiology, Smart health, regenerative and biomic medicine

1. Introduction

This chapter was inspired by and dedicated to our friend and colleague Wasyl J. Malyj, PhD (April 30, 1947–Oct. 6, 2014), shown in **Figure 1**. Wasyl was a scientist, bioinformatics pioneer, and early adopter of the central importance of heart rate variability (HRV) in health and wellness. Wasyl's achievements spanned fields as diverse as computer science, bioinformatics, genomics, nutritional science, and most relevant here, the significant role of heart rate variability in human health and performance. His role as Founding Director of the Medical Informatics Program at the University of California, Davis, and School of Medicine brought together his expertise in biomedicine, genomics, and advanced computational and network analysis skills, which he applied productively to the genomic understanding of nutritional science. We are in awe of his intellectual contributions and pay tribute to his role as a mentor and supporter of developing scientists and engineers, including ourselves. As his career matured, Wasyl became increasingly involved in developing applied technologies to harness HRV for health care research and practice, and examples of his contributions will be cited in the latter part of this chapter.

Integrative Management is a subcategory of i4P Health [1], which identifies the centrality of integrative, personalized, prescriptive, preventive, and participatory principles and practices in safe and effective health promotion, care and maintenance. Integrative Management emphasizes that the relationship between healthcare practitioner and patient is central to achieving an outcome of improved health and wellness. This is often referred to as "empowering the patient", who is seen as a central member of the treatment team. In this paper, we explore the advantages of using the integrative approach to managing chronic stress, and how new and emerging technologies clearly lead to successful outcomes in this area of health promotion. Integrative health and medicine focus on the whole person and make use of all appropriate assessment and therapeutic approaches that are informed by evidence. Integrative health care is inherently transdisciplinary. Inter-professional and traditional allopathic medicine

Figure 1.
Drs. Malyj (R) and Drury (L) exploring algorithm development from an elevated perspective.

is but one of several aspects of holistic health care. The Integrative Management framework depends on the diagnosis, treatment, and prevention of disease provided by a team of allied health professionals that includes the health-seeking individual, making optimal personal health and not just simply medical disease management, the central focus. Thus, this approach broadens the focus to include a comprehensive set of independent (diagnostic) and dependent (outcome) measures.

2. Autonomic nervous system functioning

While the important role of the autonomic nervous system (ANS) has been historically underappreciated, recently there has been a rapid proliferation of empirical, methodological, and theoretical progress in our more detailed understanding of the ANS. Previous more simplistic models of the role of the ANS using the construct of homeostasis have been enhanced by the use of the construct of allostasis and a wide variety of technological innovations including wearable and implantable biosensors have led to improved understanding of both basic and applied knowledge. This chapter will explore in particular, heart rate variability (HRV) as a rich and complex variable that has generated extensive literature, beginning with predicting all-cause mortality, but now encompassing a wide variety of disease and illness states; cognitive, affective, and behavioral processes and performance optimization. A critical analysis of HRV from the perspective of complex adaptive systems and non-linear processes will be included and innovative future uses of HRV will be described.

Normal ANS function reflects an adaptive level of interplay between the parasympathetic nervous system (PNS) and sympathetic nervous system (SNS) that produces a response to physical and psychophysiological challenge and stress. The PNS produces cardiac deceleration ("rest and digest" or "tend and befriend") and the SNS produces cardiac acceleration ("fight or flight" and stress response), while extreme stress elicits the "freeze" or "deer in the headlight" response, technically termed death feigning and can actually lead to death. However, chronic stress develops into hyper-arousal of the SNS, a process referred to as "HPA overdrive" because of the involvement of the hypothalamic-pituitary-adrenal axis. HPA overdrive causes excess glucocorticoid signaling, receptor downregulation, an end to normal negative feedback regulation of the stress response, and proliferation of peripheral pro-inflammatory cytokines by catecholamines. Thus, chronic stress reinforces more stress responding in a feed-forward cycle accompanied by a neuro-modulator presentation that is similar to depression. Psychological catastrophizing and rumination further augment the prolonged stress response and are core aspects of the expression of chronic stress. The work of Bruce McEwen of Rockefeller University and associates on allostasis and allostatic load identifies adaptive and maladaptive outcomes in the stress and coping process.

The negative health effects of chronic stress can be reduced by Autonomic Self-Regulation (ASR) because ASR dampens HPA hyper-arousal, calms the SNS, stimulates robust PNS activity, and restores normal ANS function. ASR further empowers patients to overcome the psychological sources of stress that accompany chronic nociceptive pain and self-regulate their emotions. ASR is defined as the technique of Heart Rate Variability (HRV) Biofeedback (HRVB) that incorporates (1) paced resonant frequency breathing (RFB), (2) focused attention or Mindfulness, and (3) positive emotional cognitions including those such as acceptance, compassion, gratitude, prayer, and love. ASR can rehabilitate the ANS that has been dysregulated by sensitized chronic nociceptive pain. Although HRV can be quite simply defined as a variation in the time interval between heartbeats recorded either from the ECG or a plethysmographic (pulse) sensor, this simple definition belies the complexity that exists in both the quantitative analysis of inter-beat interval (IBI) data and the fundamental systemic physiological processes that underlie HRV. Healthy HRV contains a regular pattern of increasing and decreasing IBI's between consecutive beats that increases HRV, while unhealthy HRV is relatively low due to either little variation between IBI's or random, unorganized differences between consecutive beats. In the late 1970s, low HRV was found to be a powerful clinical predictor of sudden cardiac mortality after myocardial infarction [2, 3]. By the late1980's, research revealed that adaptive cognitive performance was related to high HRV [4] and certain forms of mental disorder were related to low HRV [5].

3. Biofeedback

Biofeedback in general is simply defined as the process of gaining greater awareness of physiological functions using instruments that provide information on the activity of those same systems, with the goal of being able to control them volitionally. In addition to HR, physiological processes that can be controlled with biofeedback include electroencephalogram electromyogram and skin conductance. Clinically accurate measurement of IBI dates back to the beginning of the twentieth century, but it was the electronic digitization of computer software and increased computing power that made the quantitative analysis of HR and calculation of HRV easy and simple, accounting for the proliferation of interest in HRV. While breath or breathing training is an ancient practice with numerous forms, the production of HRV Coherence depends critically on RFB which paced breathing around 6 breaths per minute. The response of the ANS to RFB increases the amplitude of HRV rhythmic variation because 6 breath cycles/minute (= 10 seconds/cycle = 0.1 cycle/second =0.1 HZ) is the resonant frequency of the entire cardiovascular system (respiration, heart rate, baroreflex, and vasomotor tone) and parasympathetic outflow peaks [6]. Today, HRVB is widely and popularly taught and learned globally, building on the work of Gewirtz and Lehrer. Continual improvements in software algorithms and hardware have produced tools that are more efficient, more sensitive, more adaptable, more meaningful, and better visualized for collection and analysis of HRV data.

4. Heart rate variability

While HRV monitoring in the past has been done as a static 'snapshot' of HRV status with sensors cabled to a desktop or laptop computer, the future is ambulatory, real-time, and dynamic in naturalistic settings. Development of different platforms is being solicited by small business development grants by NIH/ARPA-H, DoD/DARPA, and NSF, in the private entrepreneur market, and university research to create wearable systems that are effective for reliable measurement of IBI in naturalistic environments, including home, employment, and battle settings. But many issues remain to be resolved before this movement gets going full steam into clinical practice and utility for clinical pain management: HRV does not have an accepted "gold standard" definition; not all devices have bridged from research quality to FDA approved; questions of privacy, confidentiality, and HIPAA compliance for HRV data are being confronted; third party reimbursement is poor to non-existent.

Fitness watches continuously track HR and can transfer data to a software dashboard that can compute HRV. Recent clinical research with an Apple watch app tracked people with epilepsy and found that seizures are often related to stress and missed sleep. Small chest patches with electrodes contain highly miniaturized fully-featured circuits for ECG detection. Vests are available that have HR electrode sensors and include additional sensors such as 3-axis accelerometers, respiration, skin conductance, and even more sophisticated physiological measures such as skin and ambient temperatures, "pulse-transit-time" (an indirect measure of systolic blood pressure), and EMG. The ultimate goal in signal acquisition is an entirely unobtrusive and long-lasting electrode array and signal transmission system which connects with networked devices since belts, cords, and watches are frequently not used reliably over long durations by many users. The rapidly advancing technology of biofilms with electronic monitoring capabilities will certainly assist in creating an unobtrusive, durable, and reliable HR acquisition technology. In addition to engineering fixes to this problem, the Quantitative Self Movement is making this type of assessment more culturally acceptable, with some individuals even having hardware installed in their bodies.

Ambulatory HRV monitoring has become a player in the health informatics "big data" movement. What is envisioned is having wireless transmission of HR data through processing algorithms in the cloud or through separate servers. The large-scale application of this plan falls into data mining protocols, from which new and important insights about basic HRV properties can be extracted. On the individualized level, transmitted HR data can be analyzed for comparability with normal healthy individuals and known physical and mental clinical populations as has been demonstrated by Jarczok et al. [1, 20].

Beginning in 1984, Wasyl Malyj's work anticipated the current groundswell of interest in heart rate variability and the use of biocomputation in the analysis of complex data sets. Since his initial work, the tiny literature has grown exponentially and now includes over 21,000 citations in a recent PubMed search of the term heart rate variability or HRV. Malyj's patented "Trainable adaptive focused replicator network for analyzing data" classifies signal patterns using array elements that "learn" to replicate predetermined sub-groups. This advanced wireless signal processor inputs physiological measures through large-scale databases and Malyj's patented FFT/neural network and pattern recognition algorithm. The result is fast matching of patient data to (1) provide predictive warning of acute health crises, and (2) real-time evaluation of diagnostic & treatment options for complex patient needs, using matched clinical records from other, similar cases. This is the bridge to individualized medical care plus a way to fill gaps in patient-doctor communication, as opposed to the averaged approaches dictated by today's dominant insurance/corporate models for efficient health care this, then, represents a very distinct form of personalized medicine.

Surely, one futuristic method of HR monitoring which is now a reality is remote real-time detection of pulse. Researchers have successfully deployed several different systems that measure pulse with as much accuracy as ECG: near field radar embedded in a smartphone camera programmed to display pulses as micro-movements invisible to the eye; video processing algorithm magnifying subtle changes in color reflecting redness due to pulse pressure on skin; microwave Doppler radar and more speculatively, satellite measurement of carotid artery pulsations. This is an age when the science-fictional Star Trek medical tricorder for whole-body scanning examination is no longer apocryphal (see discussion of "Berkeley Tricorder" that follows). And the acquisition, algorithmic analysis, real-time therapeutic feedback, and actionable information based on HRV are within technological reach.

Unfortunately, despite the optimistic outlook that follows from the recounting so far of opportunities for individualized and personalized integrative management of health and wellness, we must acknowledge that the United States healthcare care system suffers from significant conceptual and operational shortcomings. Theoretical and conceptual limitations of the traditional biomedical model are gradually being addressed and a fuller range of independent and dependent variables are being used, which include both individual factors and environmental issues. Because of these limitations, the United States with perhaps the highest per capita expenditure of health funds has both lower quality of outcomes and safety of the modern industrial nations. These relatively poor results come from both business practices by both the pharmaceutical and healthcare industries and inadequate governmental and regulatory These multiple factors are poorly integrated conceptually but this chapter proposes an initial synthesis of emerging technologies that can contribute useful, practical, and inexpensive indicators of health status and outcome, both for clinical and population health interventions. Such technologies can also be used for interventions such as biofeedback and patient education and self-regulation. It is axiomatic that from the scientific standpoint which can guide rational policy, without reliable and valid metrics, our understanding and ability to act is severely limited. While far from definitively addressing all of the multifarious issues of health care, a wedding of advanced technologies will catalyze progress in integrating consilient scientific knowledge. The potential role of HRV as a catalyst emerges as a practical way to improve this condition since it is a highly sensitive indicator of a wide variety of pathological conditions, diseases, and health-related phenomena. This venture is in the early development stage, but the concept has been proven and demonstrated and awaits the applications of appropriate resources to advance to operational capability.

Historically, major improvements in human health have come from public health interventions that target technological factors such as creating sanitary and salubrious environments. John Snow's removing the Broad Street pump handle terminated the London cholera epidemic and temporarily shutting down their coal-fired power plants ended a killer smog. As technology has continued to evolve, the concept of "smart technology" has emerged and led to the term Internet of Things (IoT). It is now reasonable to propose an Internet of Healthy Things (IoHT), which should be conceptualized as a public utility, rather than a consumer marketed commodity. Capra and Luisi's [7] complex adaptive systems framework can be applied to use sensor acquired bio information with networked cloud computed deep learning algorithms to produce significant improvements in health for both individuals and populations. We will now describe an emerging opportunity to use remotely acquired and network processed HRV to catalyze such a development.

5. HRV application

While the understandable major concern of most individuals is their own health status and that of their primary support network, understanding of population health and its dynamics are essential to informed policy development and practice guidelines. This is extremely timely since the current ongoing global SARS Cov 2 pandemic has devastated the health and lives of millions and the lack of monitoring and testing has been a major determinant of the unfortunate course of the multiple waves of variants of concern. Several examples of previous work that set the stage for the proposed

HRV model will be described as preconditions for a "perfect storm" of technological evolution [8, 9].

The first example is described by the National Institute of Standards as "Analysis as a Service" (AaaS) [10] and was developed by IBM [11]. Watson Analytics (WA) carries out multiple cloud-based data analyses and displays them in multiple formats as shown in **Figure 2**.

Guidi et al [12] used WA to demonstrate proof of concept for a cloud-based HRV data acquisition and analysis system which can make accurate clinical diagnostic decisions differentiating patients with Heart Failure from normal individuals on the basis of HRV. As illustrated in **Figure 3**, the process involves data acquisition using the PhysioBank and PhysioNet [13] to obtain and categorize standardized ECG data sets into the appropriate format of R to R intervals using the PhysioNet HRV Toolkit. The accuracy of prediction using HRV is displayed in **Figure 4** and compared to data from published literature, These statistics were compared to the data available in the current literature and predictive accuracy of 90% was derived. This study demonstrates proof of concept that cloud computing can generate accurate HRV data. **Figure 4** shows the results concerning accuracy of prediction using the Total Power HRV (TOT_PWR) statistic with predictive accuracy data.

The second example is the work of King et al. [14], who showed that on-scene accident triage decisions using a brief remotely obtaining HRV sample produced superior decisions to those of on-scene EMTs when requesting expensive but potentially life-saving helicopter evacuation. They used the standard deviation of non-normal intervals (SDNN), one of the candidates for broader use of HRV, and showed a sensitivity of 80% and specificity of 75%. Both the King nad Giudi studies provide proof of concept for the important role that HRV can play in healthcare settings, including triage and other diagnostic processes. It has also been demonstrated [8] that HRV can detect septicemia well before any clinical symptoms or signs emerge and that COVID-19 can be detected seven to nine days before symptoms emerge [1]. Beyond identifying pathological states and conditions, HRV has also been used to study important psychosocial functions such as executive functioning and resilience [1, 15].

As has been suggested in my previous work [8], technical developments in biosensors, microelectronics, computer networking, algorithmic data analysis such as deep learning, psychological self-regulation, and control allow a synergistic confluence which allows multiparameter continuous individual or population data

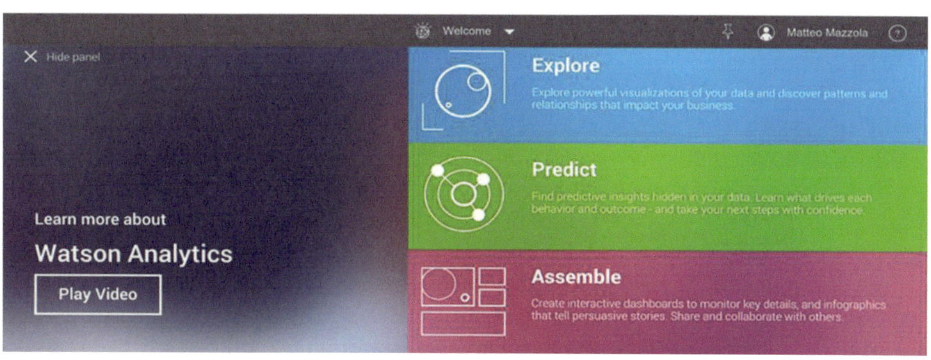

Figure 2.
IBM's Watson Analytics multiply modalities. Reproduced from Guidi et al. [12].

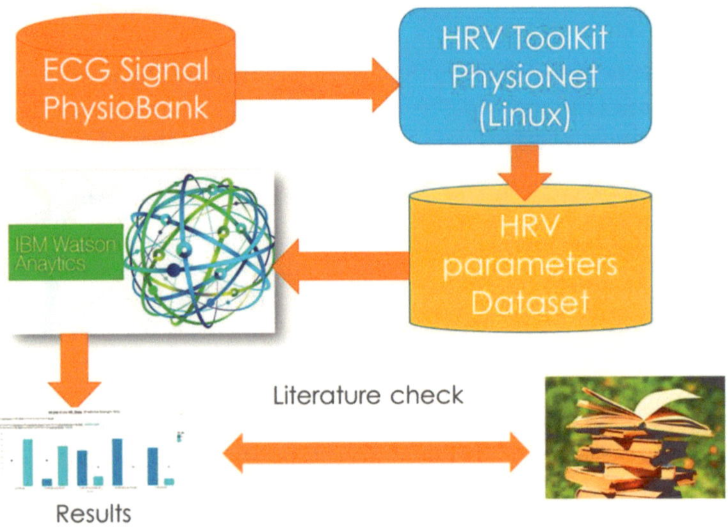

Figure 3.
Workflow diagram used in the WA process by Guidi and colleagues. Reproduced from Guidi et al. [12].

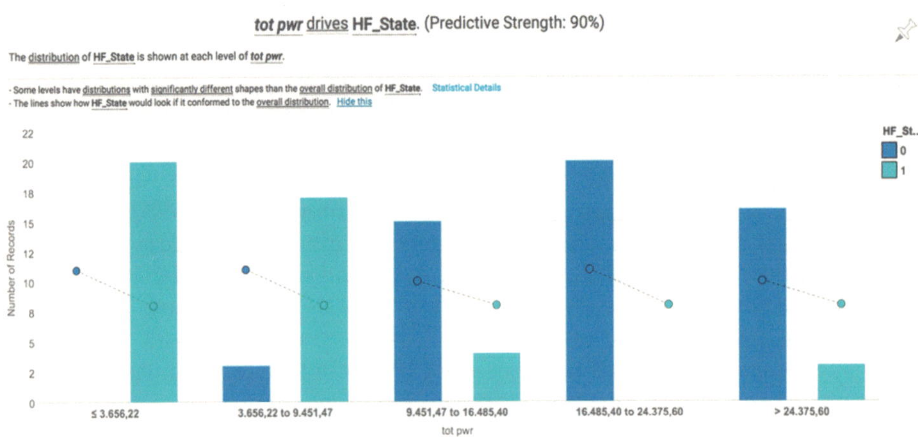

Figure 4.
WA Results using HRV Total Power as the predictor in the Guidi et al study. Reproduced from Guidi et al. [12].

for both assessment and intervention by means of a miniature electronic device. Interestingly, such a device, shown in **Figure 5**, originally dubbed by Dr. Malyj the "Berkeley Tricorder" was indeed loosely described in the prescient science fiction of Robert Heinlein in his masterpiece *Starship Troopers* and popularized in *Star Trek.* The therapeutic use of the device was described by Drury et al. [16]. The ECG data acquired can be easily analyzed by both linear and nonlinear HRV statistics. Using such devices in a networked fashion through secure encrypted data processing would realize the fictional ability Heinlein described where military team members would be in constant nonverbal communication and awareness of the functional state of each of their fellow combat team members. Such technology could easily be created to use algorithmic analysis of the aggregated HRV, respiration, and accelerometry data to indicate categorical personnel status indications: fully operational, impaired,

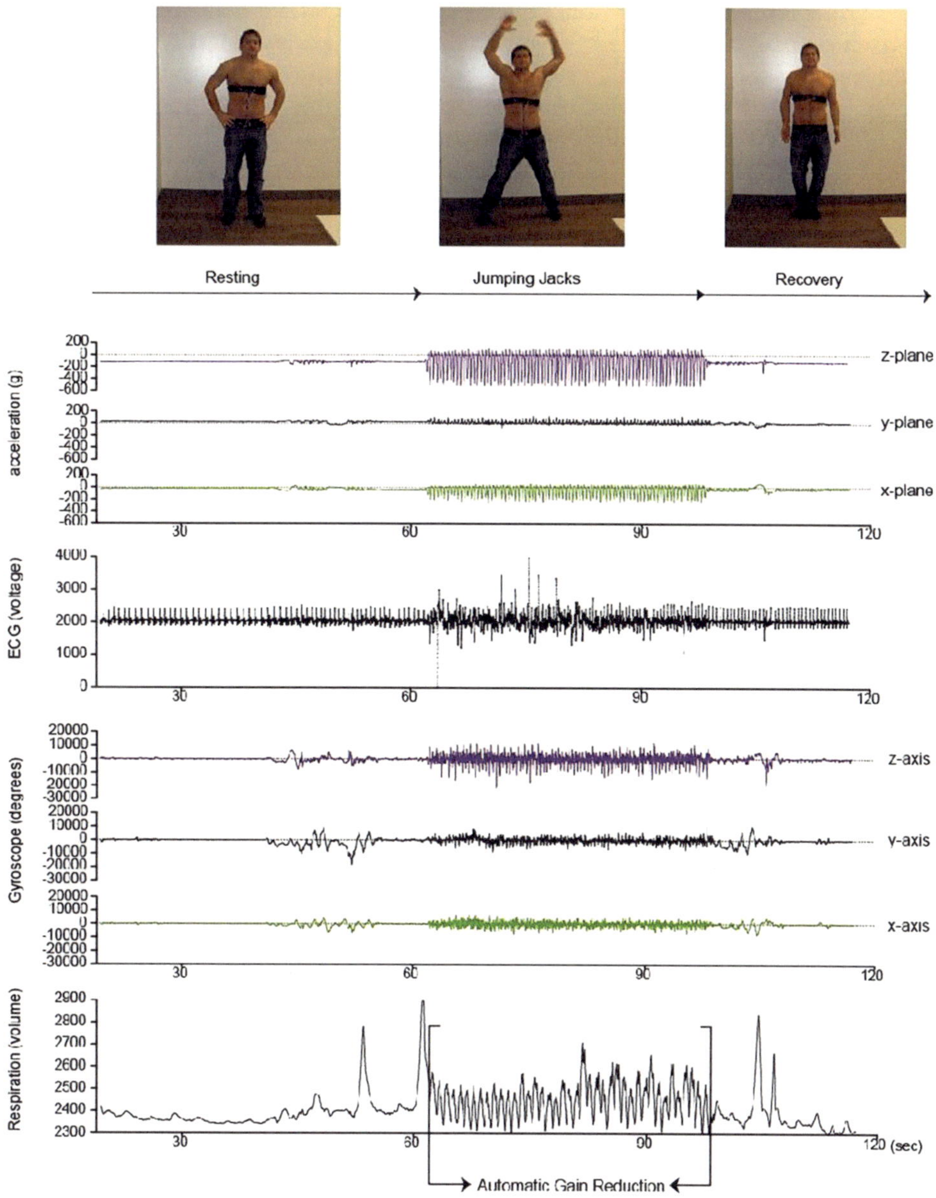

Figure 5.
Demonstrating the use of a wireless multi-parameter biosensor in conditions of rest, exercise, and recovery which was transmitted in real time to a laptop computer via Bluetooth Reproduced from Drury et al. [16].

disabled or dead, which would allow mission sensitive special operations personnel to be instantly and continuously aware of overall team functional ability and allow for fine-grained command and control on scene and at higher levels of command for decision support.

A similar technological approach could be taken to routinely monitor individual and population health status and would facilitate the early identification of deviations from healthy health parameters. Rather than waiting for the emergence of symptoms necessitating intensive, heroic, and highly expensive inpatient ICU treatment, this approach would constitute a less expensive Extensive Care System (ECS) which would blend population health, epidemiological methods with ipsative clinical intervention which could range from health promotion and disease prevention to multidimensional clinical treatment interventions. Such an application should be deemed Digital Epidemiology or Smart Health. Since this type of system would not require verbal input from patients, it could be used in assessing the health status of those who traditionally underreport symptoms, such as the elderly, and act as a check on possible overutilization because of the extensive baseline data available for individuals. Given the potential for Bluetooth enhanced bidirectional voice communication, if desired, verbal health promotion prompts and instructions could be easily delivered as well. Further, the use of smartphones with apps, which are widely used worldwide, even when little or no more conventional infrastructure exists, is practical and scalable.

This type of performance capability monitoring would also be valuable in vocational settings where fatigue and exhaustion are factors since the system described here could be enhanced with a continuous performance task, which would detect increasing signal detection errors, a sensitive measure of fatigue. The same technology could be configured to assess the ongoing ability of elders and vulnerable populations for independent living and detect the early onset of symptoms and HRV, a biomarker of disease, disability, and functional status. A similar data acquisition system using Bluetooth connection to cell devices could function as a digital epidemiology tool that would be particularly valuable in developing countries where cellphones are a primary means of communication. With the addition of EEG and EMG sensors, this device would be fully capable of conducting all-night polysomnography (PSG) in the patients' home, thereby surpassing the "gold standard" sleep laboratory PSG, with a "platinum" PSG in the natural sleep environment, eliminating the well documented "first-night effect" of sleeping in the foreign setting of a sleep laboratory, and enhancing the ecological validity of the field of polysomnography.

These uses are being facilitated and expanded by the rapid development of advanced miniaturized sensors and data acquisition materials. For example, Blaschke et al. [17] have described the use of flexible graphene-containing solution gated field-effect transistors to acquire high fidelity EEG signals in a noninvasive and unobtrusive manner. Similarly, Coleman and colleagues [18], with support of the Gates Foundation have described the use of ultra-thin stretchable and flexible devices which include adhesive peeled attachment nodes for long-term continuous monitoring of electrophysiological data. Thus, the field of advanced materials is progressing rapidly and can play an integral role in the development of iP4 Health, as can developments in genomics, data mining, cloud computing, regenerative medicine, and microbiomics [18], which have high synergistic potential.

An example of this synergistic potential is the use of HRV and other ANS techniques and concepts in the area of stem-cell and regenerative medicine. Gogolu et al. [19] summarize literature demonstrating the viability of using pluripotent human stem cells in generating enteric nervous system progenitors, while Major et al. [20] outline the step-wise differentiation of forebrain late oligodendrocyte progenitor cells (OPCs) from human pluripotent stem cells in defined chemical in vitro culture conditions. The enteric nervous system (ENS) is a key component of our enhanced view of the ANS, described as the Central Autonomic System by Thayer and Lane [21] and Benarroch [22]. Recently, Liu [23] has summarized the important and complex relationship between the microbiome, stress, and HRV. The close relationship between the nervous system, stem cell biology, and the microbiome is highly significant as an area of great importance for further research. In particular, stem cell interventions may allow modification and repair of key anatomical and physiological structures and processes.

6. Conclusion

Given the importance of global health highlighted by the Gates Foundation's Grand Challenges and others [24], a great advantage could be obtained by the use of the rapidly proliferating cellular networks that have leapfrogged traditional wired telephony and made higher computing power available through smartphones. The multi-parameter data acquisition, processing, and analysis system described above could be easily integrated into existing cellular networks and provide extensive health status monitoring in less developed and poor areas of the world. The great advantage of HRV is its high sensitivity to a very wide and diverse inventory of disorders and conditions, although it is not highly specific in identifying discrete pathology. This makes it ideal for ongoing primary health surveillance and screening in the natural environment, while follow-up evaluation is focused on specific identification and treatment of the condition. Given the digital nature of the HRV signal (interbeat intervals), it also streamlines algorithmic analysis and case identification to health personnel.

We now have the opportunity to apply new HRV technologies and algorithms in a dynamic way for a modest cost to yield powerful gains in research and development of individualized i4P health enhancement. One starting point is the use of technologies for ambulatory self-monitoring, with reliance on point-of-service medical service resources reduced, lowering costs with fewer side effects. The approaches described here represent an inflection point for translational research and development which may advance health care significantly. Despite the clearly inadequate conceptualization and deployment of the current "healthcare" system (actually a "cost containment, chronic disease management" system), the bottom line proposed here is using HRV with a suite of sister technologies as a catalyst for better health, safety and quality of life and more efficient allocation of expensive healthcare resources in an accessible manner to achieve truly smart health and wellness(IoHT).

Author details

Robert L. Drury[1,2,3]

1 ReThink Health/Canary Systems, Bainbridge Island, USA

2 Royal Medical Society, London, UK

3 Institute for Discovery, University of Wisconsin, Madison, USA

*Address all correspondence to: rl.drury@gmail.com

IntechOpen

© 2022 The Author(s). Licensee IntechOpen. This chapter is distributed under the terms of the Creative Commons Attribution License (http://creativecommons.org/licenses/by/3.0), which permits unrestricted use, distribution, and reproduction in any medium, provided the original work is properly cited.

References

[1] Drury RL. Wearable HRV biosensor system (canary) in assessing community COVID 19 prevalence. Frontiers in Autonomic Neuroscience—Horizons 2030 Special Issue on the Future of HRV. 2021;**8**:89-95. DOI: 10.3389/fnins.2021.564159

[2] Wolf M, Varigos G, Hunt D, Sloman J. Sinus arrhythmia in acute myocardial infarction. The Medical Journal of Australia. 1978;**2**(2):52-53

[3] Kleiger RE, Miller JP, Bigger JT, Moss AJ. Decreased heart rate variability and its association with increased mortality after acute myocardial infarction. American Journal of Cardiology. 1987;**59**(4):256-262

[4] Kammel H, Haase H. The visual stress model—A psycho-physiological method for the evaluation of operational reliability of pilots and cosmonauts. Acta Astronautica. 1987;**15**(2):125-132

[5] Chernigovskaya N, Vaschillo E, Petrash V, Rusanovsky V. Voluntary regulation of the heart rate as a method of functional condition correction in neurotics. Human Physiology. 1990;**16**:58-64

[6] McCraty R, Atkinson M, Tiller WA, Rein G, Watkins AD. The effects of emotions on short-term power spectrum analysis of heart rate variability. American Journal of Cardiology. 1995;**76**(14):1089-1093

[7] Capra F, Luisi P. The Systems View of Life. Cambridge: Cambridge University Press; 2014

[8] Drury R. Wearable biosensor systems and resilience: A perfect storm in health care? Frontiers in Psychology. 2014;**6**:853-860

[9] Porges S. The Polyvagal Theory. New York: Norton; 2011

[10] Mell P., Grance T, Grance T. The NIST Definition of Cloud Computing Recommendations of the National Institute of Standards and Technology. 2011. Available from: http://nvlpubs.nist.gov/nistpubs/Legacy/SP/nistspecialpublication800-145.pdf

[11] IBM Watson Analytics. A Smart Data Discovery Service Available on the Cloud, Guiding Data Exploration, Automated Predictive Analytics and Enabling Effortless Dashboard and Infographic Creation. 2017. Available from: https://www.ibm.com/analytics/watson-analytics/us-en/ [Accessed: March 5, 2017]

[12] Guidi G, Miniati R, Mazzola M, Iadanza E. Case study: IBM watson analytics cloud platform as analytics-as-a-service system for heart failure early detection. Future Internet. 2016;**8**(3):32

[13] PhysioNet. N.D. PhysioNet is Supported by the National Institute of General Medical Sciences (NIGMS) and the National Institute of Biomedical Imaging and Bioengineering (NIBIB) under NIH Grant Number 2R01GM104987-09. Available from: www.physionet.org [Accesed: March 15, 2017]

[14] King DR, Ogilvie MP, Pereira BM, Chang Y, Manning RJ, Conner JA, et al. Heart rate variability as a triage tool in patients with trauma during prehospital helicopter transport. Journal of Trauma. 2009;**67**(3):436-440

[15] Kemp AH, López SR, Passos VM, Bittencourt MS, Dantas EM, Mill JG, et al. Insulin resistance and carotid

intima-media thickness mediate the association between resting-state heart rate variability and executive function: A path modelling study. Biol. Psychology. 2016;**117**:216-224

[16] Drury R, Malyj W, Phares R. Resilience Enhancement: a psychoeducational intervention using miniaturized electronic data collection and analysis, in Presented at the American Medical Informatics Association Meeting. San Francisco: American Medical Informatics Society; 2010

[17] Blaschke BM et al. Mapping brain activity with flexible graphene micro-transistors. Frontiers in Systems Neuroscience. 2017;**4**(2):25040

[18] Kang DY, Kim YS, Ornelas G, Sinha M, Naidu K, Coleman TP. Scalable microfabrication procedures for adhesive-integrated flexible and stretchable electronic sensors. Sensors (Basel). 2015;**15**(9):23459-23476

[19] Gogolu A, Frith T, Tsakiridis A. Generating enteric nervous system progenitors fromhuman pluripotent stem cells. Current Proceedings in Stem Cell Biology. 2021;**1**(6). DOI: 10.1002/cpz1.137.

[20] Major T, Powers A, Tabar V. Derivation of telencephalic oligodendrocyte progenitors from human pluripotent stem cells. Current Proceedings in Stem Cell Biology. 2016;**39**(1):1H-10H

[21] Thayer J, Lane R. Claude Bernard and the heart–brain connection: Further elaboration of a model of neurovisceral integration. Neuroscience and Biobehavioral Reviews. 2009;**33**(2):81-88

[22] Benarroch E. The central autonomic nervous system. In: Benarroch E, editor. Autonomic Neurology. Oxford, UK: Oxford University Press; 2014. DOI: 10.1093/med/9780199920198.003.0001

[23] Liu T. The microbiome as a novel paradigm in studying stress and mental health. American Psychologist. 2016;**7**:655-667

[24] Gates Foundation. GGC. 2017. Available from: https://gcgh.grandchallenges.org [Accessed: September 1, 2017]

Chapter 8

Central Control of the Larynx in Mammals

Manuel Víctor López-González, Marta González-García, Laura Carrillo-Franco, Amelia Díaz-Casares and Marc Stefan Dawid-Milner

Abstract

Speech is a complex process that requires the coordination of multiple structures of the phonatory system regulated by the central nervous system. Specifically, the larynx is the key point necessary for the vocal folds to come into contact to convert the air that comes out of our lungs into sound. Vocal emission involves the genesis of a precise and prolonged expiration that provides an adequate pressure/air flow component to generate a subglottic pressure compatible with vocalization. The starting point for voluntary vocal production is the laryngeal motor cortex (LMC), a common structure in mammals, although the specific location within the cortex differs in humans. LCM projects to the periaqueductal gray matter (PGM), which leads to pontomedullary structures to locate the generators of laryngeal-respiratory motor patterns, necessary for vocal emission. All these regions present a high expression of FOXP2 transcription factor, necessary for brain and lung development that is closely related to vocalization. These central structures have in common that not only convey cardiorespiratory responses to environmental stress but also support vocalization. At clinical level, recent studies show that central circuits responsible for vocalization present an overactivity in certain speech disorders such as spasmodic dysphonia due to laryngeal dystonia.

Keywords: central nervous system, laryngeal motor cortex, laryngeal motoneurons, periacueductal gray matter, FOXP2, vocal emission, speech, laryngeal dystonia

1. Introduction

Central control of vocalization involves the activation of different interrelated brain structures in complex networks. Vocalization in mammals depends on a network originating in the laryngeal motor cortex, which projects to the mesencephalic Periaqueductal Gray Matter (PAG). The PAG modifies the activity of all pontomedullary structures responsible of generating all the laryngeal-respiratory motor patterns, necessary for vocal emission. These pontomedullary generators control the pattern and intensity of activation of respiratory, laryngeal, oropharyngeal, and craniofacial motor neurons [1].

Vocal emission involves the genesis of a precise and prolonged expiration that provides an adequate pressure/air flow component to generate a subglottic pressure compatible with vocalization. The nucleus ambiguus (nA), where laryngeal motor neurons are concentrated, is mainly responsible for this. All these regions present a high expression of the FOXP2 factor. FOXP2 is a transcription factor necessary for brain and lung development that is closely related to vocalization. Throughout the evolution of the human species, synaptic connectivity and plasticity in the circuits of the basal ganglia were increased, improving motor control and human cognitive and linguistic abilities [2].

Vocal fold abduction and adduction are known to be accomplished by two distinct populations of motor neurons located within the caudal third of the nA. It can be divided into three main parts: the compact formation (with motor neurons that innervate the esophagus), the semi-compact formation (with motor neurons that innervate the pharynx and the cricothyroid muscle of the larynx innervated by the superior laryngeal nerve) and the sparse formation (with motor neurons that innervate the laryngeal muscles except the cricothyroid) [3].

In previous work by our research group, the activity of the laryngeal motor neurons of nA and the reflex mechanisms involved in respiratory laryngeal responses have been characterized, suggesting that the parabrachial complex (PBc) and the A5 region (A5) have a role in modifying the activity of laryngeal motoneurones localized in the nA and accordingly the striated laryngeal muscles of the upper airway [4, 5] (**Figure 1**). Pontomedullary respiratory nuclei: PBc, A5, the nucleus of the solitary tract (NTS), nA and retroambiguous nuclei (nRA), paraambiguous (nPA) and retrofacial (nRF) integrate inputs from central and peripheral receptors and from superior structures to produce changes in the basic respiratory rhythm (eupnea). These changes are a prerequisite for survival (for example, tachypnea associated with the defense reaction, which increases the supply of oxygen preparing to fight or defend, or the response of gasping reset in the event of intense anxiety with respiratory alkalosis). But these changes in respiration are also necessary to maintain a constant

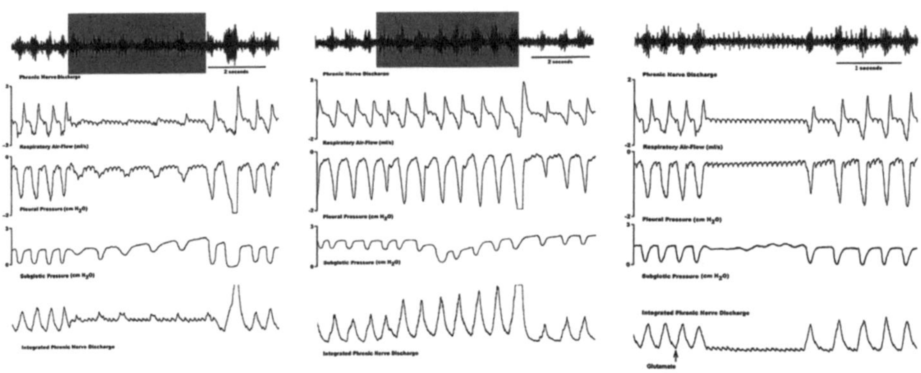

Figure 1.
Laryngeal and respiratory responses to electrical stimulation in the medial (a) and lateral (b) parabrachial nucleus and (c) to glutamate microinjection in the A5 region. Phrenic nerve discharge, respiratory airflow, pleural pressure, subglottic pressure and integrated phrenic nerve discharge showing an expiratory facilitatory response with an increase of subglottic pressure during electrical stimulation (20 mA, 0.4-ms pulses, 50 Hz for 5 s) in the medial parabrachial nucleus, an inspiratory facilitatory response with the decrease of subglottic pressure during electrical stimulation (10 mA, 0.4-ms pulses, 50 Hz for 5 s) in the lateral parabrachial nucleus and an expiratory facilitatory response with an increase of subglottic pressure during a glutamate injection (10 nl over 5 s) in the A5 region. The arrow shows the onset of injection.

expiratory flow that allows vocalization. It is known that Periaqueductal Gray Matter (PAG) is a key point in coordinating the efferent activity from limbic, cortico-prefrontal and cingulate afferents, modifying the activity of all these mesencephalic-pontomedullary nuclei [2].

2. PAG as a key point in vocalization

The PAG presents a large number of afferences. The most important have their origin from the prefrontal cortex, amygdala and hypothalamus. Its efferent projections to different pontine nuclei allow it to coordinate different patterns of cardiorespiratory and motor responses depending on the type of stimulus. Other functions of PAG include thermoregulation, participation in wakefulness and sleep mechanisms, or modulation of neuropathic pain or urination. At the clinical level, we know that its activity is modified in different neurodegenerative processes such as Alzheimer's and multisystemic atrophy [6–9].

All these higher structures that project to the PAG integrate visual, auditory and somatosensory information in the context of basic mechanisms for survival, maintaining an efferent tone on the PAG which, in turn, projects on the pontomedullary respiratory nuclei involved in respiratory rhythmogenesis to change from eupnea to a rhythm adapted to vocalization or growling. Specifically, the nRA is the perfect target to convert passive breathing into active breathing to generate motor activities that produce changes in abdominal pressure, in addition to modifying the activity of the motor neurons that are located in the nA and that control the caliber of the pharynx and larynx [10, 11]. Stimulation of PAG and nRA is known to produce vocalization [12] and lesions in PAG cause mutism in animals and humans [13, 14] and vocalization and problems in the production of voice when lesions occur in nRA [15]. However, the electrophysiological influence of PAG on these pontomedullary nuclei has not yet been described.

3. Vocalization in apes: PAG-laryngeal motor cortex connectivity

Regarding the studies of the pathways that participate in voluntary and involuntary vocalizations, there is a model that explains vocal control that includes two hierarchically organized pathways. Involuntary vocalizations are innate and require a different control mechanism than that which dominates voluntary vocalizations or speech [16, 17]. These emotional expressions, such as crying or laughing, are directed by the emotional system, made up of specific pathways that target the brain stem and spinal cord [18, 19]. Specifically, research carried out with the squirrel monkey has determined that the system includes: the cingulate gyrus, the PAG, and various pontine and medullary nuclei [20–24].

The PAG receives projections from the upper limbic regions and from cortical areas such as the anterior cingulate gyrus, insula, and orbitofrontal cortex. In addition, it maintains connections with the caudal part of the nRA. The nRA has direct access to the motor neurons involved in vocalization, that is, it controls the motor neuron groups that innervate the soft palate, pharynx, and larynx, as well as the diaphragm, intercostal, abdominal, and pelvic muscles. Its final objective is to control/modify the intra-abdominal, intrathoracic and subglottic pressure, the control of which is essential to generate vocalization.

In primates, vocalizations, in addition to activation of the PAG, can be produced by electrical stimulation of the hypothalamus [20, 25, 26], amygdala [20], bed nucleus of the stria terminalis [22], orbitofrontal cortex [27] and anterior cingulate gyrus [28], since all these regions have a strong connection with the PAG [29–33]. The only necessary condition is that the PAG was intact [13, 34, 35].

On the contrary, if areas that are not connected with the PAG, such as the motor or premotor cortex, are stimulated, no vocalizations are produced [36, 37]. These results emphasize the essential role of the PAG in the production of vocalization in primates as well as in humans. This coordinating role in the generation of vocalization is also demonstrated by the fact that the activation of the caudal levels of the PAG can generate partial vocalizations through its connection with the nRA [10, 38, 39].

In summary, the input of the stimulus occurs in the primary integrating center of vocalizations (VOC). Next, the superior temporal gyrus, the supplementary motor area and the insula, will be in charge of modulating the stimulus. Once the output from the VOC is produced, the stimulus can be directed to the corticobulbar pathway and cerebellum directly or, on the contrary, it can go from the cingulate gyrus, the PAG, the pons, until reaching the reticular area of the medulla; which has access to the nA ipsilaterally and contralaterally [40]. On the other hand, the voluntary production of voice in human beings consists of a sound modulation of sound. This production depends directly on the laryngeal motor cortex, that is, the production of voluntary vocal emissions in humans, requires the activation of this cortex, located in the dorsal portion of the ventral zone of the primary motor cortex, and its direct connection with the laryngeal motor neurons of the nA, which are in charge of controlling the laryngeal muscles for the emission of learned vocal patterns [41].

However, it has been shown that during speech emissions there is a joint activation of the voluntary and involuntary system [19]. An involuntary activation of the path takes place automatically to give the vocal emissions the appropriate emotional character. Therefore, stimulation of the pathway that originates in the primary motor cortex and runs through the PAG and nRA is required; in addition to the activation of the pathway that goes from the laryngeal motor cortex directly to the corticomedullary fibers, which will activate the motor neurons of the face, mouth, tongue, larynx and pharynx, to control the production of words and phrases.

Finally, vocal control will depend on the primary motor area, a bilateral structure that is responsible for laryngeal control and orofacial musculature [42], in addition to the activation of the superior temporal gyrus to compensate for alterations in the auditory feedback of the tone used during phonation. Likewise, two feedback loops are put into operation that provide the motor cortex with the information necessary to carry out motor commands for phonation. One of these loops includes the basal ganglia, while the other involves the cerebellum. However, these structures seem unnecessary for the production of innate vocal patterns [43, 44].

Therefore, we see that emotional emissions in humans require bilateral activation of the laryngeal motor cortex (**Figure 2**) [45]. Furthermore, the system for the production of speech involves a predominant activation of the left hemisphere, including the superior temporal gyrus, the anterior insula, the basal ganglia, and the cerebellum. For this production, the activity of the cingulate gyrus and the PAG is also necessary, in variable degrees, to associate the emotional character with the vocal production.

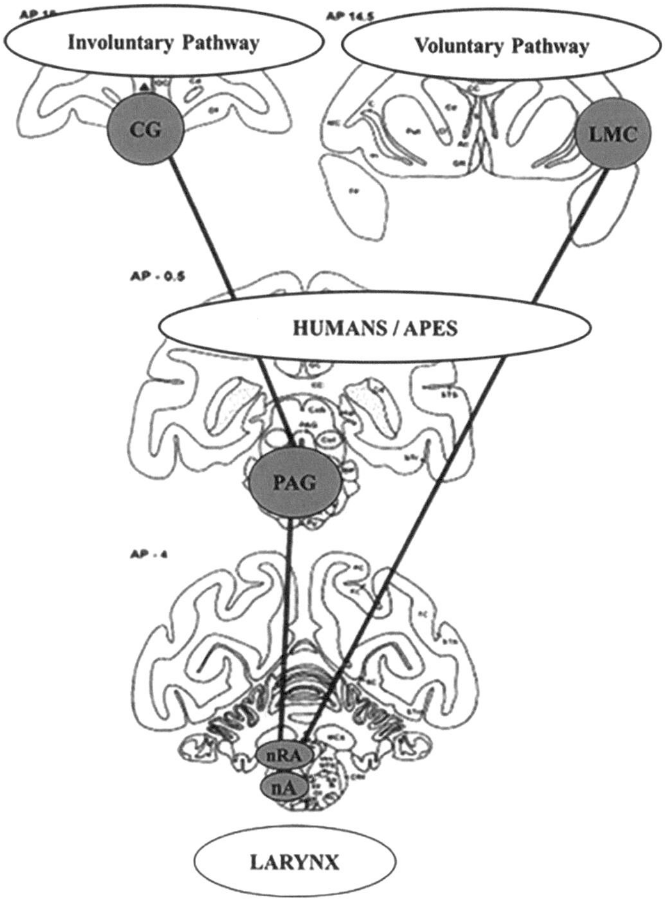

Figure 2.
Scheme of squirrel monkey brain coronal section showing the voluntary and involuntary pathways controlling the larynx.

4. Other structures functionally related to the PAG

Other intermediate structures project both to the laryngeal motor cortex and the nA. One of these structures is PBc, cited above [46]. Neuronal unit recording experiments demonstrate the presence of neuronal activity during vocalization [47], suggesting that this nucleus is involved in laryngeal motor coordination as an intermediate nucleus of proprioceptive information between the cortex and the nA. In addition, recent work has shown that both cPB and PAG have high immunoreactivity to the expression of the FOXP2 gene, demonstrating that both regions are primarily involved in modulating the expiratory flow necessary for the production of the sound and voice [2].

In one of the latest works carried out by our research group, we have demonstrated that the column corresponding to the dorsolateral PAG is involved in the control of defense response [48, 49], which is associated with tachycardia, hypertension and redistribution of blood flow. This sympathetic response is mediated by the rostroventrolateral medulla (RVLM), which, in turn, activates sympathetic preganglionic neurons present in the intermediolateral column of the spinal cord. These projections are ultimately responsible for the sudden increase in blood pressure. It is also known that the increase in blood pressure is produced by indirect activation of the RVLM by other less studied pathways [50–52].

The dorsolateral PAG does not have direct connections with the RVLM but has very dense connections (afferents and efferents), with the hypothalamic area responsible for the activation of the RVLM during the defense response. We know that the cardiorespiratory activity of dorsolateral PAG neurons depends on the activity of these hypothalamic neurons [52, 53]. Likewise, our group has shown that there are functional connections from this hypothalamic area and other pons structures such as the parabrachial complex (cPB) [54, 55] and A5 area [56, 57]. Both regions are rich in FOXP2 expression. Our research group has shown the importance of the interrelation between some of these hypothalamic-midbrain and pontomedullary structures involved in cardiorespiratory control. More specifically, we have focused on the analysis of these interactions by analyzing the defense response evoked from specific areas of the hypothalamus (dorsomedial hypothalamic area and perifornical area (DMH-PeF)) [54–57] and midbrain (PAG) (**Figures 3** and **4**) [48, 49].

Originally, the stimulation of these areas evokes a series of cardiorespiratory and autonomic changes that characterize the defense response [58]. The defense response prepares the animal for situations environmental stresses that require a rapid locomotor response characterized hemodynamically by hypertension, tachycardia and redistribution of blood flow from abdominal and visceral areas to the skeletal muscles of the extremities. Additionally, this response is accompanied by mydriasis, increases in respiratory rate and tidal volume and vocalization [56].

Recent publications propose DMH-PeF as one of the main areas of the hypothalamus that generate the defense response carried by the PAG [59]. Disinhibition of DMH-PeF after microinjection of bicuculin (GABA receptor antagonist) produces an increase in renal sympathetic activity and blood pressure that has been attributed to the activation of neurons in RVLM [60, 61]. The administration of bicuculin also produces an increase in heart rate, which decreases between 30 and 50% due to the inhibition of Rafe Pallidus with muscimol [62–64]. In addition, there is morphological evidence of projections from the DMH to the Rafe Pallidus [65]. The results seem to show that the pressor and tachycardia cardiovascular responses, typical of stress, evoked from this region, including in the DMH-PeF, would present two descending

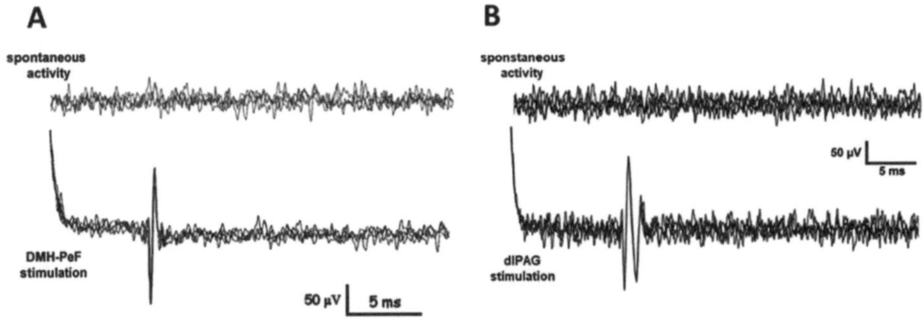

Figure 3.
Extracellular recordings of two putative cells were recorded from the A5 region. (A) Silent neuron (upper trace, 4 superimposed sweeps) with constant-latency responses to HDA stimulation (lower trace). The cell was demonstrated to be orthodromically activated from the HDA. (B) Silent neuron (upper trace, 4 superimposed sweeps). The lower trace shows constant latency responses (4 superimposed sweeps) to the dlPAG stimulation.

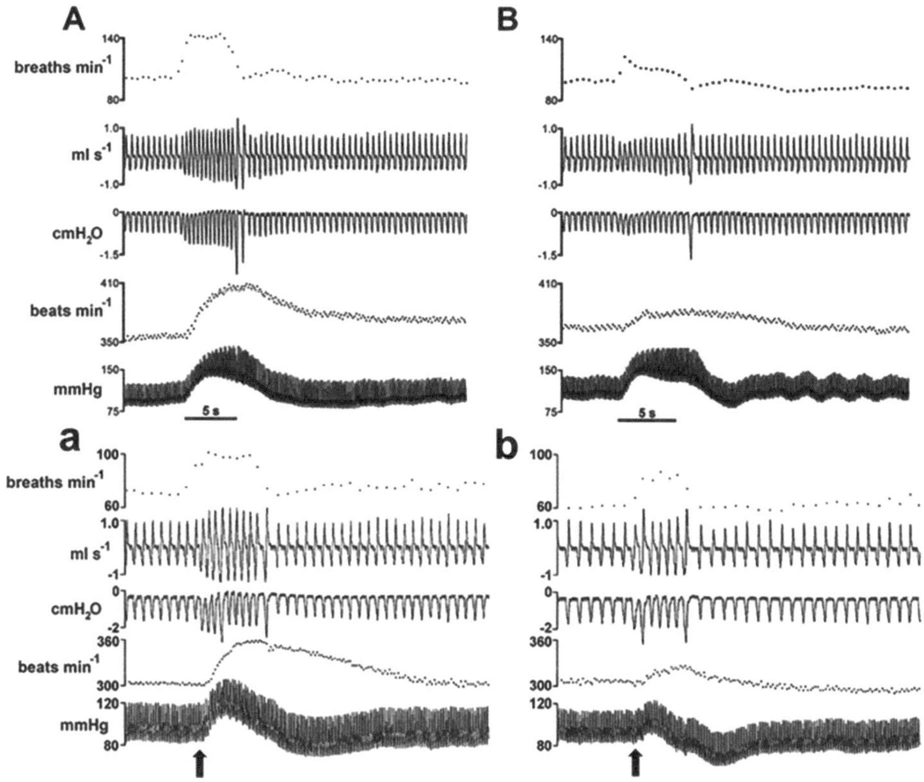

Figure 4.
Instantaneous respiratory rate (upper trace), respiratory flow, pleural pressure, instantaneous heart rate and blood pressure in a spontaneously breathing rat, showing the cardiorespiratory response evoked on PAG (upper figures) or HDA (lower figures) stimulation before [A, a] and after [B, b] the microinjection of muscimol (50 nl over 5 s) in the A5 region. The arrows indicate show the onset of the HDA/PAG electrical stimulation.

routes, one responsible for the increase in blood pressure, via RVLM, and the other responsible for the increase in heart rate, via Pale Rallidus. Both responses use the PAG as an intermediate station [60].

5. Preliminary results

In previous studies, we have characterized the activity of laryngeal nA motor neurons and the reflex mechanisms involved in respiratory laryngeal responses. We have also described the existence of a network of hypothalamic-midbrain-pontomedullary nuclei that modulate the cardiorespiratory responses produced to certain types of stress. Bearing this in mind and knowing that many of these structures express FOXP2 and participate in vocalization processes, the limited number of existing publications that study the electrophysiological relationships between the neural circuits involved in the control of laryngeal activity is striking.

Recently, we have been able to carry out a series of preliminary experiments with the techniques that we had been using for years and that we are taking up again. The results of this previous approach include the "in vivo" recording of laryngeal motor neurons and the recording of subglottic pressure and laryngeal resistance using the technique of "isolated glottis in situ" and the analysis of the changes that occur in these parameters. During electrical stimulation of the study areas. These works have led to two publications in international congress proceedings; [66] in Proc Physiol Soc 43, PC208 and [67] in J Physiol Biochem, 74 (Suppl 1) and two communications presented at the SENC Congress in Santiago de Compostela (2019), concluding that not also PAG but CnF seem to modify the activity of laryngeal motoneurons.

6. Clinical implications

Therefore, all these central structures described have in common the fact that they convey cardiorespiratory responses to environmental stress and support vocalization. Recent studies show that the laryngeal microstructure and its innervation undergo the same changes during development in rodents and humans [68] and that the central circuits responsible for vocalization present an overactivity in certain speech disorders of central origin such as spasmodic dysphonia due to laryngeal dystonia [69].

Laryngeal respiratory apnea is frequently a particularly serious clinical manifestation, as occurs in newborn apnea or central sleep apneas, caused by immaturity or abnormalities of central respiratory control in these individuals, causing an exaggerated response of the respiratory system, the laryngeal adduction reflex [70].

Furthermore, it is also known that spasmodic dysphonia, a focal form of dystonia, is a neurological alteration of the voice that manifests with involuntary "spasms" of the vocal cords, which result in speech interruptions and affect the quality of the voice. The two recognized types of spasmodic dysphonia are adductor spasmodic dysphonia (intermittent excessive closure of the vocal cords) and abductor spasmodic dysphonia (prolonged opening of the vocal cords). The cause of spasmodic dysphonia is unknown, although there is some consensus that behind it there is an alteration of the central nervous system, especially at the level of motor control. Specifically, alterations have been described in the circuit of the ganglia of the base, cerebellum and sensorimotor cortex, and structural alterations in the corticobulbar and corticospinal tracts, which are the nerve tracts that come into contact with the bulbar neurons responsible for phonation [69]. A better description and knowledge of the individual contribution of each of the nuclei that make up this hypothalamic-midbrain network on the central control of laryngeal motor neurons, would allow a better understanding not only of normal phonatory control, but would also contribute to a better understanding of the central alterations produced in this type of dystonia as well as in other disorders at the vocal level.

Paradoxical laryngeal adduction movements are characterized by adduction or approximation of the vocal cords during the respiratory cycle (especially during the inspiratory phase), which causes airway obstruction at the laryngeal level. The resulting dyspnoea and stridor are frequently confused with asthma, but do not respond to treatment with steroids and bronchodilators, since the glottic narrowing is independent of the caliber of the bronchial lumen. The origin of this intermittent interruption of transglottic airflow due to paradoxical laryngeal adduction remains to be elucidated. It has been linked to laryngeal irritation from agents such as gastroesophageal reflux or acute severe stress [71].

On the other hand, the pressor response and tachycardia associated with the stimulation of the DMH-PeF and PAG, mediated by the cPB nuclei and A5 Area, are of immediate interest for the knowledge of certain types of hypertensions classified clinically as "essential". We believe that knowledge of the mechanisms involved in the inhibition of the baroreceptor reflex, which could mediate this type of hypertension, will allow us to provide new data that will contribute to better explaining the mechanisms of neuronal interactions between the hypothalamic-midbrain regions and the pontomedullary cardiorespiratory and laryngeal control centers implicated in this type of pathology.

7. Summary and perspectives

During the last two decades, Health Sciences research has evolved from a purely biological perspective towards a biopsychosocial model of health and disease. As a result, it has been found that there is a relationship between voice disorders and neuro-vegetative responses associated with emotional responses, mainly those related to anxiety and stress. These responses are generated by the activation of the hypothalamic defense areas and are carried by the PAG and pontomedullary structures such as the PBc and A5. The emotional response intervenes, along with other psychological factors, on the tone of the laryngeal muscles causing spasmodic dysphonia or laryngeal dysphonia. This occurs because the laryngeal muscles appear to be extremely sensitive to emotional stress generated by anxiety, anger, irritability, impatience, frustration, and depression, which can lead to spasmodic dysphonia or laryngeal dystonia [69]. Along these lines, Demmink-Geertman et al. [72] confirmed that, due to the characteristics of the higher pitch of the female voice, this effect is greater in women of all ages and that, above all, it affects professionals who use the voice as a means of work. This fundamentally affects women involved in teaching tasks. Only in Andalusia, the number of teachers is 132,985, and in Spain, they exceed 750,000, of which 71.9% are women. The percentage is particularly relevant in early childhood (97.6%), special (81.7%) and primary (81.4%) education. At least 21% have vocal involvement and 15.8% of sick leave is due to voice problems (FETE-UGT 2019 teaching report). Knowing the pathophysiology of the mechanisms by which stress produces alterations in the functionality of the vocal cords would allow the development of adequate treatments for these pathological processes.

Therefore, new contributions are needed to add new perspectives to a series of pathologies that are related to mechanisms that have their origin in these hypothalamic-midbrain regions, such as the so-called central apneas associated with hypertension [70], apneas associated with sudden death infantile syndrome [73], paradoxical laryngeal adduction movements [71] and muscular tension dysphonia secondary to stress [74].

Acknowledgements

The study was supported by a program grant Junta de Andalucía, Group n° CTS-156, Spain.

Conflict of interest

The authors declare that they have no conflict of interest.

Author details

Manuel Víctor López-González[1,2,3*], Marta González-García[1,2,3], Laura Carrillo-Franco[1,3], Amelia Díaz-Casares[1,2,3] and Marc Stefan Dawid-Milner[1,2,3]

1 Faculty of Medicine, Department of Human Physiology, University of Málaga, Málaga, Spain

2 Unit of Neurophysiology of the Autonomic Nervous System (CIMES), University of Málaga, Málaga, Spain

3 Biomedical Research Institute of Málaga (IBIMA), Spain

*Address all correspondence to: manuelvictor@uma.es

IntechOpen

© 2022 The Author(s). Licensee IntechOpen. This chapter is distributed under the terms of the Creative Commons Attribution License (http://creativecommons.org/licenses/by/3.0), which permits unrestricted use, distribution, and reproduction in any medium, provided the original work is properly cited.

References

[1] Ludlow CL. Central nervous system control of interactions between vocalization and respiration in mammals. Head Neck. 2011;**33** Suppl 1:S21-5. DOI: 10.1002/hed.21904

[2] Stanic D, Dhingra R, Dutchmann M. Expression of the transcription factor FOXP2 in brainstem respiratory circuits of adult rat is restricted to upper airway premotor areas. Respiratory Physiology & Neurobiology. 2018;**250**:14-18. DOI: 10.1016/j.resp.2018.01.014

[3] Pascual-Font A, Hernández-Morato I, McHanwell S, Vázquez T, Maranillo E, Sañudo J, et al. The central projections of the laryngeal nerves in the rat. Journal of Anatomy. 2011;**219**(2):217-228. DOI: 10.1111/j.1469-7580.2011.01390.x

[4] Dawid-Milner MS, Lara JP, Milán A, González-Varón S. Activity of inspiratory neurones of the ambiguus complex during cough in the spontaneously breathing decerebrate cat. Experimental Physiology. 1993;**78**(6):835-838. DOI: 10.1113/expphysiol.1993.sp003730

[5] Lara JP, Dawid-Milner MS, Lopez MV, Montes C, Spyer KM, Gonzalez-Baron S. Laryngeal effects of stimulation of rostral and ventral pons in the anaesthetized rat. Brain Research. 2002;**934**(2):97-106. DOI: 10.1016/s0006-8993(02)02364-8

[6] Behbehani MM. Functional characteristics of the midbrain periaqueductal gray. Progress in Neurobiology. 1995;**46**:575-605. DOI: 10.1016/0301-0082(95)00009-k

[7] Carrive P. The periaqueductal gray and defensive behavior: Functional representation and neuronal organization. Behavioural Brain Research. 1993;**58**:27-47. DOI: 10.1016/0166-4328(93)90088-8

[8] Heinricher MM, Tavares I, Leith JL, Lumb BM. Descending control of nociception: Specificity, recruitment and plasticity. Brain Research Reviews. 2009;**60**:214-225. DOI: 10.1016/j.brainresrev.2008.12.009

[9] Zhang W, Hayward LF, Davenport PW. Respiratory responses elicited by rostral versus caudal dorsal periaqueductal gray stimulation in rats. Autonomic Neuroscience: Basic and Clinical. 2007;**134**:45-54. DOI: 10.1016/j.autneu.2007.02.003

[10] Subramanian HH, Holstege G. The nucleus retroambiguus control of respiration. The Journal of Neuroscience. 2009;**29**:3824-3832. DOI: 10.1523/jneurosci.0607-09.2009

[11] Boers J, Klop EM, Hulsshoff AC, De Weerd H, Holstege G. Direct projections from the nucleus retroambiguus to cricothyroid motoneurons in the cat. Neuroscience Letters. 2002;**319**:5-8. DOI: 10.1016/s0304-3940(01)02395-3

[12] Holstege G, Kerstens L, Moes MC, Vanderhorst VG. Evidence for a periaqueductal gray-nucleus retroambiguus-spinal cord pathway in the rat. Neuroscience. 1997;**80**:587-598. DOI: 10.1016/s0306-4522(97)00061-4

[13] Jürgens U, Pratt R. Role of the periaqueductal grey in vocal expression of emotion. Brain Research. 1979;**167**:367-378. DOI: 10.1016/0006-8993(79)90830-8

[14] Esposito A, Demeurisse G, Alberti B, Fabbro F. Complete mutism after midbrain periaqueductal gray

lesion. Neuroreport. 1999;**10**:681-685. DOI: 10.1097/00001756-199903170-00004

[15] Shibba K, Umezaki T, Zheng Y, Miller AD. The nucleus retroambigualis controls laryngeal muscle activity during vocalization in the cat. Experimental Brain Research. 1997;**115**:513-519. DOI: 10.1007/pl00005721

[16] Jürgens U. Neural pathways underlying vocal control. Neuroscience and Biobehavioral Reviews. 2002; **26**:235-258. DOI: 10.1016/s0149-7634 (01)00068-9

[17] Jurgens U. The neural control of vocalization in mammals: A review. Journal of Voice. 2009;**23**:1-10. DOI: 10.1016/j.jvoice.2007.07.005

[18] Holstege G. The emotional motor system. European Journal of Morphology. 1992;**30**:67-79

[19] Holstege G, Bandler R, Saper CB. The emotional motor system. Progress in Brain Research. 1996;**107**:3-6. DOI: 10.1016/s0079-6123(08)61855-5

[20] Jürgens U, Maurus M, Ploog D, Winter P. Vocalization in the squirrel monkey (Saimiri sciureus) elicited by brain stimulation. Experimental Brain Research. 1977;**4**:114-117. DOI: 10.1007/ BF00240356

[21] Jürgens U, Ploog D. Cerebral representation of vocalization in the squirrel monkey. Experimental Brain Research. 1970;**10**:532-554. DOI: 10.1007/ BF00234269

[22] Jürgens U. Projections from the cortical larynx area in the squirrel monkey. Experimental Brain Research. 1976;**25**:401-411. DOI: 10.1007/ BF00241730

[23] Muller-Preuss P, Jürgens U. Projections from the 'cingular' vocalization area in the squirrel monkey. Brain Research. 1976;**103**:29-43. DOI: 10.1016/0006-8993(76)90684-3

[24] Hage SR, Jurgens U. Telemetric recording of neuronal activity. Methods. 2006;**38**(3):195-201. DOI: 10.1016/j. ymeth.2005.08.005

[25] Baxter BL. Comparison of the behavioral effects of electrical or chemical stimulation applied at the same brain loci. Experimental Neurology. 1967;**19**:412-432. DOI: 10.1016/ 0014-4886(67)90162-8

[26] Altafullah I, Shipley C, Buchwald JS. Voiced calls evoked by hypothalamic stimulation in the cat. Experimental Brain Research. 1988;**71**:21-32. DOI: 10.1007/BF00247519

[27] Myers RE. Comparative neurology of vocalization and speech: Proof of a dichotomy. Annals of the New York Academy of Sciences. 1976;**280**:745-760. DOI: 10.1111/j.1749-6632.1976.tb25537.x

[28] Hunsperger RW, Bucher VM. Affective behaviour by electrical stimulation in the forebain and brain stem of the cat. Progress in Brain Research. 1967;**27**:103-127. DOI: 10.1016/ s0079-6123(08)63095-2

[29] Hopkins DA, Holstege G. Amygdaloid projections to themes encephalon, pons and medulla oblongata in the cat. Experimental Brain Research. 1978;**32**:529-547. DOI: 10.1007/ BF00239551

[30] Holstege G, Meiners L, Tan K. Projections of the bed nucleus of the stria terminalis to the mesencephalon, pons, and medulla oblongata in the cat. Experimental Brain Research.

1985;**58**:379-391. DOI: 10.1007/BF00235319

[31] Holstege G. Anatomical evidence for an ipsilateral rubrospinal pathway and for direct rubrospinal projections to motoneurons in the cat. Neuroscience Letters. 1987a;**74**:269-274. DOI: 10.1016/0304-3940(87)90308-9

[32] Holstege G. Descending motor pathways and the spinal motor system: Limbic and non-limbic components. Progress in Brain Research. 1991;**87**:307-421. DOI: 10.1016/s0079-6123(08)63057-5

[33] Kuipers R, Mensinga GM, Boers J, Klop EM, Holstege G. Infralimbic cortex projects to all parts of the pontine and medullary lateral tegmental field in cat. The European Journal of Neuroscience. 2006;**23**:3014-3024. DOI: 10.1111/j.1460-9568.2006.04843.x

[34] Adametz J, O'Leary JL. Experimental mutism resulting from periaqueductal lesions in cats. Neurology. 1959;**9**:636-642. DOI: 10.1212/wnl.9.10.636

[35] Skultety FM. Experimental mutism in dogs. Archives of Neurology. 1962;**6**:235-241. DOI: 10.1001/archneur.1962.00450210063007

[36] Jürgens U. On the elicitability of vocalization from the cortical larynx area. Brain Research. 1974;**8**:564-566. DOI: 10.1016/0006-8993(74)90853-1

[37] Kirzinger A, Jürgens U. Cortical lesion effects and vocalization in the squirrel monkey. Brain Research. 1982;**233**:299-315. DOI: 10.1016/0006-8993(82)91204-5

[38] Schadt CR, Cox KL, Tramontana MG, Byrne DW, Davis TL, Fang JY, et al. Depression and intelligence in patients with Parkinson's disease and deep brain stimulation. Journal of the National Medical Association. 2006;**98**:1121-1125

[39] Subramanian HH, Arun M, Silburn PA, Holstege G. Motor organization of positive and negative emotional vocalization in the cat midbrain periaqueductal gray. The Journal of Comparative Neurology. 2015;**524**(8):1540-1557. DOI: 10.1002/cne.23869

[40] Ludlow CL. Central nervous system control of voice and swallowing. Journal of Clinical Neurophysiology. 2015;**32**(4):294-303. DOI: 10.1097/WNP.0000000000000186

[41] Simonyan K. The laryngeal motor cortex: Its organization and connectivity. Current Opinion in Neurobiology. 2014;**28**:15-21. DOI: 10.1016/j.conb.2014.05.006

[42] Brown S, Ngan E, Liotti M. A larynx area in the human motor cortex. Cerebral Cortex. 2008;**18**:837-845. DOI: 10.1093/cercor/bhm131

[43] Larson CR, Burnett TA, Kiran S, Hain TC. Effects of pitch-shift velocity on voice Fo responses. Journal of the Acoustical Society of America. 2000;**107**:559-564. DOI: 10.1121/1.428323

[44] Loucks TM, Poletto CJ, Simonyan K, Reynolds CL, Ludlow CL. Human brain activation during phonation and exhalation: Common volitional control for two upper airway functions. NeuroImage. 2007;**36**:131-143. DOI: 10.1016/j.neuroimage.2007.01.049

[45] Penfield W, Roberts L. Speech and Brain Mechanisms. Princeton, N.J: Princeton University Press; 1959

[46] Geerling J, Yokota S, Rukhadze I, Roe D, Chamberlin N. Kölliker–Fuse

GABAergic and glutamatergic neurons project to distinct targets. The Journal of Comparative Neurology. 2017;**525**:1844-1860. DOI: 10.1002/cne.24164

[47] Farley GR, Barlow SM, Netsell R. Factors influencing neural activity in parabrachial regions during cat vocalizations. Experimental Brain Research. 1992;**89**(2):341-351. DOI: 10.1007/BF00228250

[48] Peinado-Aragonés CA. Interrelaciones de la sustancia gris periacueductal dorsolateral y la región protuberancial a5 en el control central cardiorrespiratorio (tesis doctoral). Málaga: Universidad de Málaga; 2016

[49] López-González MV, González-García M, Peinado-Aragonés CA, Barbancho MA, Díaz-Casares A, Dawid-Milner MS. Pontine A5 region modulation of the cardiorespiratory response evoked from the midbrain dorsolateral periaqueductal grey. Journal of Physiology and Biochemistry. 2020;**76**(4):561-572. DOI: 10.1007/s13105-020-00761-1

[50] Keay KA, Bandler R. Parallel circuits mediating distinct emotional coping reactions to different types of stress. Neuroscience and Biobehavioral Reviews. 2001;**25**:669-678. DOI: 10.1016/s0149-7634(01)00049-5

[51] Hayward LF. Midbrain modulation of the cardiac baroreflex involves excitation of lateral parabrachial neurons in the rat. Brain Research. 2007;**1145**:117-127. DOI: 10.1016/j.brainres.2007.01.140

[52] De Menezes RC, Zaretsky DV, Fontes MAP, DiMicco JA. Cardiovascular and thermal responses evoked from the periaqueductal grey require neuronal activity in the hypothalamus. The Journal of Physiology. 2009;**587**:1201-1215. DOI: 10.1113/jphysiol.2008.161463

[53] Horiuchi J, McDowall LM, Dampney RA. Vasomotor and respiratory responses evoked from the dorsolateral periaqueductal grey are mediated by the dorsomedial hypothalamus. The Journal of Physiology. 2009;**587**:5149-5162. DOI: 10.1113/jphysiol.2009.179739

[54] Díaz-Casares A, López-González MV, Peinado-Aragonés CA, Lara JP, González-Barón S, Dawid-Milner MS. Role of the parabrachial complex in the cardiorespiratory response evoked from hypothalamic defense area stimulation in the anesthetized rat. Brain Research. 2009;**1279**:58-70. DOI: 10.1016/j.brainres.2009.02.085

[55] Díaz-Casares A, López-González MV, Peinado-Aragonés C, González-Barón S, Dawid-Milner MS. Parabrachial complex glutamate receptors modulate the cardiorespiratory response evoked from hypothalamic defense area. Autonomic Neuroscience. 2012;**169**(2):124-134. DOI: 10.1016/j.autneu.2012.06.001

[56] López-González MV, Díaz-Casares A, Peinado-Aragonés CA, Lara JP, Barbancho MA, Dawid-Milner MS. Neurons of the A5 region are required for the tachycardia evoked by electrical stimulation of the hypothalamic defence area in anaesthetized rats. Experimental Physiology. 2013;**98**:1279-1294. DOI: 10.1113/expphysiol.2013.072538

[57] López-González MV, Díaz-Casares A, González-García M, Peinado-Aragonés CA, Barbancho MA, Carrillo de Albornoz M, et al. Glutamate receptors of the A5 region modulate cardiovascular responses evoked from the dorsomedial hypothalamic nucleus and perifornical area. Journal of Physiology and Biochemistry. 2018;**74**(2):325-334. DOI: 10.1007/s13105-018-0623-3

[58] Mancia G, Zancchetti A. Hypothalamic control of autonomic

functions. In: Morgane PJ, Panksepp J, editors. Handbook of the Hypothalamus. Behavioral Studies of the Hypothalamus. New York: Dekker; 1981. pp. 147-201

[59] Dampney RAL. Central neural control of the cardiovascular system: Current perspectives. Advances in Physiology Education. 2016;**40**(3):283-296. DOI: 10.1152/advan.00027.2016

[60] Fontes MAP, Tagawa T, Polson JW, Cavanagh SJ, Dampney RAL. Descending pathways mediating cardiovascular response from dorsomedial hypothalamic nucleus. The American Journal of Physiology. 2001;**280**:H2891-H2901. DOI: 10.1152/ajpheart.2001.280.6.H2891

[61] Cao WH, Fan W, Morrison SF. Medullary pathways mediating specific sympathetic responses to activation of dorsomedial hypothalamus. Neuroscience. 2004;**126**:220-240. DOI: 10.1016/j.neuroscience.2004.03.013

[62] Samuels BC, Zaretsky DV, DiMicco JA. Tachycardia evoked by disinhibition of the dorsomedial hypothalamus is mediated through medullary raphe. The Journal of Physiology. 2002;**538**:941-946. DOI: 10.1113/jphysiol.2001.013302

[63] Zaretsky DV, Zaretskaia MV, DiMicco JA. Stimulation and blockade of GABA(A) receptors in the raphe pallidus: Effects on body temperatura, heart rate, and blood pressure in conscious rats. American Journal of Physiology. Regulatory, Integrative and Comparative Physiology. 2003;**285**(1):R110-R116. DOI: 10.1152/ajpregu.00016.2003

[64] Horiuchi J, McAllen RM, Allen AM, Killinger S, Fontes MA, Dampney RA. Descending vasomotor pathways from the dorsomedial hypothalamic nucleus: Role of medullary raphe and RVLM. American Journal of Physiology. Regulatory, Integrative and Comparative Physiology. 2004;**287**:R824-R832. DOI: 10.1152/ajpregu.00221.2004

[65] Samuels BC, Zaretsky DV, DiMicco JA. Dorsomedial hypothalamic sites where disinhibition evokes tachycardia correlate with location of raphe-projecting neurons in rats. American Journal of Physiology-Regulatory, Integrative and Comparative Physiology. 2004;**287**:R472-R478. DOI: 10.1152/ajpregu.00667.2003

[66] González-García M, López-González MV, Carrillo-Franco L, Carrillo-Franco C, Dawid-Milner MS. Hypothalamic and mesencephalic regions involved in the control of laryngeal activity and subglottic pressure in spontaneously breathing anaesthetized rats. Proceedings of the Physiological Society. 2019;**43**:PC208

[67] González-García M, López-González MV, Díaz-Casares A, Peinado-Aragonés CA, Barbancho MA, Dawid-Milner MS. Does the midbrain dorsolateral periaqueductal grey have direct connections with the pontine a5 region? A neuropharmacologic and electrophysiological study. FENS Abstr. 2018;**9**:E004

[68] Peterson JR, Watts CR, Morris JA, Shelton JM, Cooper BG. Laryngeal aging and acoustic changes in male rat ultrasonic vocalizations. Developmental Psychobiology. 2013;**55**(8):818-828. DOI: 10.1002/dev.21072

[69] Mor N, Simonyan K, Blitzer A. Central voice production and pathophysiology of spasmodic dysphonia. The Laryngoscope. 2018;**128**:177-183. DOI: 10.1002/lary.26655

[70] Drager LF, Genta PR, Pedrosa RP, Nerbass FB, Gonzaga CC, Krieger EM. Characteristics and predictors of

obstructive sleep apnea in patients with systemic hypertension. The American Journal of Cardiology. 2010;**105**(8):1135-1139. DOI: 10.1016/j.amjcard.2009.12.017

[71] Christopher KL, Morris MJ. Vocal cord dysfunction, paradoxic vocal fold motion, or laryngomalacia? Our understanding requires an interdisciplinary approach. Otolaryngologic Clinics of North America. 2010;**43**(1):43-66, viii. DOI: 10.1016/j.otc.2009.12.002

[72] Demmink-Geertman L, Dejonckere PH. Differential effects of voice therapies on neurovegetative symptoms and complaints. Journal of Voice. 2010;**24**(5):585-591. DOI: 10.1016/j.jvoice.2008.12.013

[73] Wang X, Guo R, Zhao W, Pilowsky P. Medullary mediation of the laryngeal adductor réflex: A posible role in sudden infant síndrome. Respiratory Physiology and Neurobiology. 2016;**226**:121-127. DOI: 10.1016/j.resp.2016.01.002

[74] Van Houtte E, Van Lierde K, Claeys S. Pathophysiology and treatment of muscle tension dysphonia: A review of the current knowledge. Journal of Voice. 2011;**25**:2. DOI: 10.1016/j.jvoice.2009.10.009

Chapter 9

Signaling Pathways Regulating Axogenesis and Dendritogenesis in Sympathetic Neurons

Vidya Chandrasekaran

Abstract

The post-ganglionic sympathetic neurons play an important role in modulating visceral functions and maintaining homeostasis through complex and reproducible axonal and dendritic connections between individual neurons and with their target tissues. Disruptions in these connections and in sympathetic nervous system function are observed in several neurological, cardiac and immune-related disorders, which underscores the need for understanding the mechanisms underlying neuronal polarity, axonal growth and dendritic growth in these neurons. The goals of this chapter are to explore our current understanding of the various growth factors, their signaling pathways, downstream effectors and interplay between these pathways to regulate different stages of axonal and dendritic growth in sympathetic neurons.

Keywords: sympathetic neurons, growth factors, neurotrophins, cytokines, BMPs, axons, dendrites

1. Introduction

The sympathetic nervous system is an important component of the peripheral autonomic nervous system responsible for controlling the visceral functions of the body to maintain homeostasis and the "flight or fight response" [1]. The sympathetic pathway is composed of two neurons – a preganglionic neurons located in the intermediolateral horn of the spinal cord, originating from the thoracolumbar region of the spinal cord and the postganglionic neuron that is, in most cases, located in the paravertebral sympathetic ganglia chain on either side of the spinal cord. Some of the preganglionic axons synapse with pre-vertebral sympathetic ganglia such as the celiac, mesenteric and pelvic ganglia, which innervate the gastrointestinal and urinary tracts and are not part of the sympathetic chain [2]. The superior cervical ganglia (SCG) is the first and the largest ganglia in the sympathetic chain and innervates most of the tissues in the head and neck region including the pineal gland, cerebral blood vessels, carotid body, vestibular system, muscles in the iris, lacrimal glands and piloerector muscles. Of the sympathetic neurons, SCG neurons are one of the most studied to understand various aspects of neuronal development in the peripheral nervous system. In recent years, the observation of autonomic dysfunction in many diseases such

IntechOpen

as Parkinson's disease, cardiac disorders, multiple system atrophy, multiple sclerosis, diabetes and immune-related disorders, has renewed an interest in understanding neuronal development and maintenance of sympathetic neurons [3–11].

During early development, the precursors of the post-ganglionic sympathetic neurons are derived from the trunk neural crest cells, which then migrate ventrally along the neural tube, through the anterior portion of the sclerotome and coalesce near the dorsal aorta to form the sympathetic ganglia [12]. In rodents, the neural crest migration occurs between E8 and E11, with cells forming coalesced sympathetic ganglia around E12–E14 with the more rostral ganglia forming before the caudal ones. Studies on the early sympathetic neuron specification and neural crest migration show that growth factors such as neurotrophins, semaphorins and ephrins are important for migration of these neural crest cells, with bone morphogenetic proteins (BMPs) being important for their differentiation into sympathetic neuronal lineage. The exposure to BMPs leads to the induction o of transcription factors such as Phox2b, Mash1, Hand2, Gata3, Insm1, Sox4 and Sox 11, which lead to the survival of these neurons and their differentiation into noradrenergic neurons [12]. Following the specification of these neurons, the next crucial step to create a functional sympathetic network is the extension and maturation of axons and dendrites. In this chapter, we will explore the pathways that are important for establishing and refining axonal and dendritic arbors in sympathetic neurons.

2. Growth factors and signaling pathways involved in axonal growth

Following the specification of sympathetic neurons, the first sign of neuronal polarity is the extension of a single axon from the cell body [13]. In rodents, although the initiation of axonal growth from sympathetic ganglia starts as early as E12, most of the axonal growth occurs around E14–E15, with target innervation continuing into first few weeks of postnatal life [13–15]. Axonal growth has three stages – initiation of axons from the post-ganglionic neurons, elongation of the axons towards the final targets and finally target innervation which involves branching as well as restriction of axonal growth. Research using cultured sympathetic neurons *in vitro* and *in vivo* studies have identified multiple growth factors, extracellular matrix molecules and downstream signaling targets, involved in different stages of axonal growth and axonal guidance, functioning either as activators or inhibitors of axonal growth. In this section, we will examine the various molecules and their roles in these three stages of axonal growth.

2.1 Hepatocyte growth factor

Hepatocyte growth factor (HGF) or scatter factor is one of the few growth factors that appears to be involved in initiation of axonal growth in sympathetic neurons. Both HGF and its receptor Met tyrosine kinase are co-expressed in the sympathetic neurons throughout embryonic, starting as early as E12.5, with HGF being secreted by the sympathetic neurons and functioning as an autocrine regulator of axonal growth [16–18]. Treatment of cultured sympathetic neurons with HGF induces axonal growth and enhances the axonal growth promoted by nerve growth factor [16]. Also, inhibition of HGF activity through treatment with anti-HGF antibodies and *Met* signaling mutants show decreased axonal growth and branching compared to wildtype embryos [17–19]. Although HGF promotes survival of sympathetic neuroblasts, it is not necessary for the survival of post-mitotic sympathetic neurons [17].

Furthermore, *in vitro* studies and studies on docking site mutants for Met receptors suggest that HGF exerts its effects on axonal growth in mice through activation of the mitogen-activated protein kinase (MAPK) pathway and PI-3 K pathway [20]. Although lack of HGF signaling *in vivo* results in decreased axonal growth, it does not lead complete lack of axons in sympathetic neurons, suggesting the involvement of other factors in the first step of axonal growth.

2.2 Artemin

Artemin, a member of the glial derived neurotrophic factor (GDNF) family ligands (GFLs) plays an important role in the axonal elongation and guidance of the postganglionic axons to their targets [21–24]. In addition to Artemin, other members of the GDNF family, including GDNF and Nerturin have been shown to enhance neurite growth in subpopulations of sympathetic neurons [21, 22, 24]. Artemin mRNA is expressed at high levels near the dorsal aorta around E12.5 and then in the smooth muscles of many of the blood vessels along which the sympathetic axons migrate to their targets [25, 26]. The receptors for Artemin – Ret and GFRa3 are both expressed in the sympathetic ganglia as early as E11.5 and then expression gets restricted to subsets of cells later in embryonic development [26–29]. Treatment of nascent sympathetic ganglia (E13.5) with artemin induces axonal growth with axons showing branching and radial outgrowth. Also, axonal growth from explant cultures of the ganglia are directed towards beads coated with artemin, suggesting artemin has the ability to guide axons to their targets [30]. In addition, Artemin knockout mice show decreased axonal growth postnatally, [31] and mice lacking either GFRa3 or Ret show reduced, depleted or abnormal neuronal projections and abnormal branching indicating that Artemin signaling mediated by Ret:GFRa3 receptor complex is necessary for proper migration of sympathetic neurons during development [26, 28]. Although early studies suggest a role for Artemin in sympathetic neuron survival with the superior cervical ganglia being smaller in Artemin, Ret and GFRα3 knockout animals compared to wild type animals [29, 32], more recent studies suggest that the decreased neuronal cell numbers in the absence of Artemin signaling are an indirect effect of aberrant axonal migration and target innervation [28]. Taken together, the data suggest that the members of the GDNF family act as early guidance molecules to promote axon elongation and target innervation.

2.3 Neurotrophins

Neurotrophin family of growth factors – nerve growth factor (NGF), neurotrophin-3 (NT-3), neurotrophin-4 (NT-4) and brain derived neurotrophic factor (BDNF) have been implicated in many aspects of neuronal function, including differentiation, survival, axonal growth and dendritic growth [33–37]. These neurotrophins are synthesized and secreted as proneurotrophins, which are the proteolytically cleaved to generate mature neurotrophins that activate different isoforms of Trk tyrosine kinase receptors (Trk) and p75 neurotrophin receptors ($p75^{NTR}$) and activate a variety of downstream signaling pathways [33–36, 38].

2.3.1 Nerve growth factor

NGF synthesis begins in targets of sympathetic neurons in concert with the arrival of the sympathetic axons and is correlated with increased expression of

TrkA [39–43]. Although NGF is necessary for survival for sympathetic neurons in the early stages, the neurons lose their dependence on NGF for survival in the later stages *in vivo* and *in vitro* [42, 44, 45]. Exposure of sympathetic neurons to exogenous NGF, overexpression of NGF in the target tissues or adding NGF to compartments containing the distal axons leads to increased axonal growth and hyperinnervation of the target tissues [46–51]. Conversely, mice lacking NGF or TrkA (NGF receptor) show decreased survival of sympathetic neurons and decreased target innervation [52–54]. In addition to NGF, proNGF promotes axonal elongation and branching in postnatal sympathetic neurons through activation of the $p75^{NTR}$ receptor rather than the TrkA receptor [55].

NGF's axonal growth effects are independent of its effects on neuronal survival. Mice lacking both NGF and Bax, a pro-apoptotic gene necessary for apoptosis in sympathetic neurons [56, 57], show normal early axonal growth and guidance along the vasculature but show differential loss of innervations in the different target tissues with sympathetic innervation being completely absent in salivary glands and cardiac ventricles, reduced in the liver and unaffected in the trachea [58]. These evidence supports the argument that NGF is important for axon growth of the distal axons and target innervation. It is interesting to note that this requirement for target-derived NGF in the terminal axonal growth varies between the different targets, suggesting that other growth factors are important for target innervation in some of these tissue [58].

NGF's effects on axonal growth are primarily mediated through activation of the TrkA receptors. NGF and phosphorylated TrkA are retrogradely transported in endosomes from the axon terminals [59–62] and regulate axonal growth through changes to cytoskeletal proteins and transcription factors such as cyclic AMP response element binding protein (CREB) and early growth regulator 3 (Egr3) [49, 63–67]. Also, local reintroduction of NGF to NGF-deprived neurons in culture results in profuse axonal growth, suggesting that NGF promotes axonal growth both locally and through retrograde signaling [68]. NGF also upregulates the expression of its receptor TrkA in sympathetic neurons [41] and activates downstream effectors such as PI-3Kinase-Akt pathways and MAPK pathways leading to cytoskeletal changes resulting in axonal growth [69]. In addition, NGF, through its binding to TrkA receptors, activates glycogen synthase kinase-3 (GSK-3), which results in the phosphorylation of microtubule-associated protein 1B (MAP1B) and decrease in MAP1B phosphorylation is correlated with decreased axonal growth [70]. NGF signaling during axonal elongation and termination is dependent on activation SHP-2, a protein tyrosine phosphatase. Inhibition of SHP-2 *in vitro* leads to decreased axonal growth by inhibiting extracellular signal-regulated kinase (ERK) signaling, however interfering with SHP-2 signaling results in increased axonal density within the targets [71]. Also, studies suggest that Wnt 5a is upregulated in sympathetic targets in response to NGF, and blocking Wnt5a activation using an antibody suppresses NGF-induced axonal growth [72]. Early growth response (Egr) proteins – Egr1 and Egr3 are induced by NGF signaling in sympathetic neurons with inhibition of Egr1 *in vitro* using a dominant negative and Egr3 knockout *in vivo* show decreased neurite outgrowth and target innervation [49, 73, 74]. Recent studies have suggest a role of non-coding RNAs and post-translational modifications downstream of NGF signaling during axonal growth [75, 76]. Untranslated axonal mRNA Tp53inp2 upregulates NGF-TrkA signaling during axonal growth [75] and NGF-dependent prenylation of proteins such as Rac GTPase appears to be important for receptor trafficking to promote axonal growth [76].

Once the axons reach the target, NGF-TrkA signaling increases the expression of Coronin-1, a protein that interacts with the actin cytoskeleton [77]. Coronin-1 acts as a molecular switch to convert downstream effectors of NGF-TrkA from the PI-3 K pathway to calcium signaling, leading to the suppression of axonal growth and branching [77, 78].

2.3.2 Neurotrophin 3 (NT-3)

In addition to NGF, neurotrophin-3 is expressed in sympathetic neurons, although its main receptor TrkC is expressed at low levels in neonatal sympathetic neurons [44, 79–81]. NT-3 mutant mice show severe defects in their sympathetic nervous system with 50% fewer neurons, and defects in axonal branching and axonal innervation of target tissues such as the pineal gland and cardiac myocytes [82–84]. In addition, neurotrophin-3 (NT-3) promotes axonal growth and branching in sympathetic neurons *in vitro* [41, 84]. Overexpression of NT-3 in adipose tissue leads to increased sympathetic innervation through its activation of TrkC receptors [85]. However, NT-3's effects on axonal growth are mediated by activation of TrkA receptors as opposed to TrkC, with NT-3 selectively promoting neurite outgrowth rather than for survival in neonatal sympathetic neurons [41]. Although both NGF and NT-3 signal using the same receptor, unlike the NGF-TrkA complex, NT-3-TrkA complex does not mediate retrograde signaling [61]. Recent studies also suggest that NT3-TrkA complex prevents axons from branching into intermediate targets and enables larger growth cones in the absence of Coronin-1, through activation of Ras-MAPK and PI3K-Akt pathways [86].

2.3.3 Brain-derived neurotrophic factor (BDNF)

Similar to other neurotrophins, BDNF is expressed in sympathetic neurons and sympathetic neuron targets [79, 87], and serves as target-derived growth factor for pre-ganglionic sympathetic neurons [88]. Unlike NGF and NT-3, BDNF null mutants show a slight increase in the number of sympathetic neurons compared to wildtype animals, indicating that BDNF is not important for survival of sympathetic neurons [89]. Addition of exogenous BDNF inhibits axonal growth and inhibiting BDNF activity using antibodies against BDNF promotes axonal growth in sympathetic neurons *in vitro* [79]. Also, BDNF $^{+/-}$ and BDNF $^{-/-}$ mice show hyperinnervation of the target tissues [87]. Although Trk B (the main BDNF receptor) is not present in sympathetic neurons [41], the sympathetic axons express $p75^{NTR}$ during target innervation [87] and BDNF's effects on axonal growth are mediated through its interaction with this receptor. BDNF-$p75^{NTR}$ signaling inhibits the activity of NGF-TrkA complex leading to axonal growth inhibition *in vitro* and axon pruning *in vivo* [87, 90].

2.4 Tumor necrosis factor superfamily

Multiple members of the tumor necrosis factor superfamily (TNFSF) are known to regulate axonal growth in sympathetic neurons. Members of the TNFSF act as either as membrane-bound ligands or soluble ligands once cleaved from the membrane and bind to receptors belonging to the TNF superfamily (TNFRSF) [91, 92]. These molecules can also serve as reverse signaling molecules with TNFRSF acting as ligands and membrane-bound TNFSF functioning as receptors [93].

TNFa protein is present in postnatal SCG neurons throughout the cell body and neurites with strong immunoreactivity for TNF receptors R1 (TNFR1) in the cell body and in target tissues [94]. *tnfa*$^{-/-}$ and *tnfr*$^{-/-}$ mice show decreased innervation of sympathetic targets, with no effect on neuronal numbers [94]. While soluble TNFa inhibits NGF-induced axonal growth *in vitro* through activation of NF-kB [95], the reverse signaling mediated by TNFR1 at the axon terminal enhances axonal growth and target innervation through elevation of opening of T-type calcium channels leading to rapid activation of protein kinase C, ERK1 and ERK2 [94, 96]. Another TNF superfamily member – receptor-activator of NF-κB (RANK, also known as TNFRSF11A)) is expressed in embryonic and early postnatal sympathetic neurons, while its ligand RANKL is expressed in target tissues [97]. Similar to TNFa, local activation of RANKL-RANK signaling is necessary for axonal growth effects, and addition of soluble RANKL or activation of RANK signaling inhibits NGF-induced axonal extension and branching, through activation of NF-κB signaling [97]. The glucocorticoid induced tumor necrosis factor receptor related protein (GITR) and its ligand GITRL are also expressed in sympathetic neurons [98]. The activation of GITR by its ligand GITRL leads to activation of ERK signaling and the downregulation of NF-kB signaling pathways and regulation of both of these pathways are necessary for NGF-induced axonal growth [98, 99]. A recent study showed that *TWE-PRIL*, an alternative spliced form that combines extracellular domain of one TNFSF member APRIL (TNFSF13) and the transmembrane and cytoplasmic domains of another member TWEAK (TNFSF12), is expressed in developing SCG neurons [100]. *April* $^{-/-}$ mice show increased axonal growth in the presence of NGF, that can be rescued by overexpression of TWE-PRIL. TWE-PRIL reverse signaling leads to axonal growth inhibition by preventing NGF-dependent activation of ERK [100]. Similarly, CD40 (TNFRSF5) and its ligand CD40L are expressed in embryonic and early postnatal SCG neurons [101]. While CD40 by itself does not affect axonal growth, the reverse autocrine signaling mediated by CD40-CD40L enhances NGF induced axonal growth in these neurons, especially when there is low NGF with high levels of NGF inhibiting CD40 and CD40L expression [102].

Interestingly, two TNF family members have differential effects on paravertebral and prevertebral ganglia. Unlike SCG targets which showed hypoinnervation in *tnfa*$^{-/-}$ and *tnfr*$^{-/-}$ mice, the targets of the prevertebral sympathetic ganglia showed no change in innervation and reverse signaling mediated by TNFR1 did not alter axonal growth from prevertebral ganglia neurons [103]. Similarly, CD40 null mutants show hyperinnervation in targets of prevertebral ganglia and CD40-CD40L reverse signaling inhibits axonal growth in prevertebral ganglia neurons [101].

2.5 Extracellular matrix proteins

As axons extend from the sympathetic ganglia to the target, they are exposed to a complex environment composed of extracellular matrix molecules such as laminin, collagen, fibronectin and thrombospondin. Laminin, collagen IV and thrombospondin promote axonal growth in perinatal superior cervical ganglia neurons *in vitro* and mediate their effects on axonal growth through activation of specific classes of integrin receptors [104–107]. Exposure of SCG neurons to laminin leads to formation of multiple axons, whereas neurons exposed to collagen IV extend only a single axon suggesting distinct signaling pathways downstream of integrin activation [104, 105, 108]. In addition, exposure to laminin causes bundling of microtubules, leading to rapidly growing axons [109]. Conversely, chondroitin sulfate proteoglycans inhibit axonal growth in cultured neonatal SCG neurons and may be responsible for lack of sympathetic reinnervation of the heart during ischemia-reperfusion injury [110].

2.6 Other signaling pathways involved in axonal growth

Interleukin 1b (IL-1b) and Interleukin 1 receptor (IL-1R) are expressed in neonatal sympathetic neurons with IL-1R1 being present in the cell body and axons, and IL-1b being expressed in the sympathetic neurons and target tissues [111, 112]. IL-1b inhibits axonal growth in cultured sympathetic neurons by promoting the nuclear translocation of NF-kB [112].

Ceramide, a lipid second messenger, generated from glycosphingolipid metabolism or sphingomyelin metabolism is known to be important for cell proliferation or cell death downstream of extracellular agents such as TNF, interleukins and other molecules [113, 114]. Although newly synthesized glycosphingolipids are not important for axonal growth, when added to the distal axons ceramide inhibits neuronal outgrowth, possibly by decreasing the uptake of NGF by the distal axons [113, 114].

3. Growth factors and signaling pathways involved in dendritic growth regulation

Dendritogenesis in post-ganglionic sympathetic neurons begins around E14, with maturation of dendritic arbor continuing into postnatal development [13, 115]. Sympathetic neurons extend multiple dendrites with complex branching patterns. The size of the dendritic arbor is dependent on size of the target field and neuronal activity, suggesting that dendritic complexity is determined by the needs of the targets [116–120]. Similar to axonal growth, dendritogenesis can be divided into 3 stages – initiation of dendrites, elongation and branching of dendrites, and maturation coupled with pruning of the dendritic tree. In this section, we will explore the current understanding of the various growth factors, their signaling pathways and interactions between them to influence dendritic arborization in sympathetic neurons.

3.1 Bone morphogenetic proteins (BMPs)

Members of the bone morphogenetic protein (BMP) family are important for dendritic growth initiation in sympathetic neurons *in vitro* and *in vivo*. BMPs bind and activate a heterotrimeric receptor complex of transmembrane serine/threonine kinase receptors made of type I receptor – BMP receptor type I A (BMPR1a), also known as activin receptor-like kinase-3 (ALK-3) or BMP receptor type IB (BMPR1b), also known as ALK-6, and one of the three type II receptors – BMPRII, Activin type II or IIB (Act II or ActRIIB). The activation of these kinases leads to phosphorylation of receptor Smads (Smads 1, 5 and/or 8), which complex with Smad 4 to translocate to the nucleus and regulate gene expression [121–123].

Sympathetic neurons and glial cells in the SCG from embryonic and postnatal ganglia express mRNA and protein for BMP-5, BMP-6 and BMP-7 [108, 124, 125]. Also, BMPR1a, BMPRIIB, ActRII and BMPRII are present in mouse SCG through later stages of embryonic development into postnatal life [126, 127], suggesting that BMP signaling pathway is functional in sympathetic neurons during periods of dendritogenesis. BMP-5, BMP-6, BMP-7 initiate dendritic growth in cultured perinatal sympathetic neurons by activation and translocation of the Smad complex and regulating gene expression [108, 128, 129]. Conditional knockouts of BMPR1a or both BMPR1a/1b show a decrease in dendritic length and branch complexity compared to congenic wildtype animals but do not show complete absence of dendrites [130]. Also, BMP receptor knockouts showed a dramatic decrease in total dendritic length, branching and soma size later in postnatal development suggesting that BMP signaling may be important for maintenance of dendrites, rather than initiation of dendrites *in vivo* [130]. This difference in BMP function *in vitro* and *in vivo* may stem from the presence of other receptors such as the activin receptors to mediate BMP signaling [131]. Interestingly, although transfection of Smad1 dominant negative mutant blocks BMP-7-induced dendritic growth *in vitro*, the SCG neurons in conditional Smad 4 knockout mice show an increase in dendritic length and total dendritic arbor [130], suggesting that Smad 4 may play a limiting role *in vivo* and BMPs may be signaling through Smad-dependent and Smad-independent pathways for dendritic growth regulation.

Transcriptome and miRNome analyses have identified over 250 genes and over 40 microRNAs whose expression are altered in response to BMP-7 treatment in cultured sympathetic neurons during the period of dendritic growth initiation [132, 133]. Of the genes, $p75^{NTR}$ mRNA and protein are strongly upregulated by BMP-7 signaling in cultured SCG neurons. BMP-mediated effects on dendritic growth are not observed in $p75^{NTR}$ knockout mice, with $p75^{NTR}$ knockout mice showing stunted dendritic arbor compared to wildtype. Conversely, overexpression of $p75^{NTR}$ phenocopies the dendritic growth effects of BMP-7, suggesting that this is an important target of BMP-7 during dendritic growth regulation [132, 134]. However, $p75^{NTR}$ ligands, interplay between neurotrophins and BMP in activating $p75^{NTR}$ and downstream effectors of $p75^{NTR}$ signaling responsible for dendritogenesis in sympathetic neurons still need to be elucidated. Of the microRNAs identified, three miRNAs – miR-21, miR-23b and miR-664-1* may regulate dendritic growth downstream of BMP-7 in sympathetic neurons *in vitro* [133]. Also, signaling pathways mediated by ubiquitin-proteasome system and by reactive oxygen species are suggested to be downstream of Smad signaling in sympathetic neurons *in vitro* with proteasome inhibitors and antioxidants inhibiting BMP-7 induced dendritic growth [135, 136].

3.2 Neuronal activity dependent dendritic growth

Electric field stimulation or treatment of sympathetic neurons with potassium chloride can lead to neuronal depolarization and this neuronal activity triggers dendritic growth in postganglionic sympathetic neurons by the activation of calcium calmodulin dependent kinase II (CaMKII) [137]. Also, inhibition of integrin-linked kinase (ILK) using an siRNA prevents activity-dependent dendritic growth in sympathetic neurons *in vitro*, whereas pharmacological inhibition of glycogen synthase kinase-3β (GSK-3β) enhances activity-dependent dendritic growth in these neurons [138, 139].

3.3 Nerve growth factor and fibroblast growth factor

NGF was one of the earliest growth factors recognized as important for dendritic growth with NGF injections leading to enhanced dendritic growth in sympathetic ganglia [116]. However, NGF, by itself, is unable to induce dendritic growth in cultured perinatal SCG neurons, but is required for BMP-7 induced dendritic growth [129, 140]. One of the downstream targets of NGF for dendritic growth appears to be Egr3 with Egr3−/− mice showing significant decrease in the number of primary dendrites, total dendritic length and maximum extent of dendritic arbor [74].

DOI: http://dx.doi.org/10.5772/intechopen.102442

Fibroblast growth factor receptor 1 (FGFR1) is expressed in adult SCG neurons and its nuclear localization increases in perinatal sympathetic neurons upon BMP-7 exposure [141, 142]. Also, expression of mutant FGFR1 decreases the dendritic growth induced by BMP-7 in sympathetic neurons, through the activation of the integrative nuclear FGFR1 signaling pathway [142].

Interestingly, stimulation of the MAPK signaling pathways has differential effects on activity-dependent dendritic growth and BMP-7 induced dendritic growth. While pharmacological inhibition of ERK activity using PD98059 inhibits activity-dependent dendritic growth, the treatment with the same inhibitor enhances BMP-7-induced dendritic growth [137, 139, 143]. Stimulation of the MAPK signaling through overexpression of MEK1 leads to inhibition of BMP-7 induced dendritic growth and the inhibition of MAPK signaling pathway with dominant negative MEK1 or ERK2 mutant increases the number of dendrites and total dendritic arbor in BMP-7 treated [143]. Further studies are needed to understand the opposing roles of ERK in BMP-induced vs. activity-dependent dendritic growth.

3.4 Cytokines

Several members of the cytokine family have been shown to regulate dendritic growth in sympathetic neurons. These growth factors function through the activation of the Janus kinase (JAK), leading to the nuclear translocation of proteins known as signal transducers and activators of transcription (STAT) [144]. In perinatal sympathetic neurons, interferons gamma (IFNg), leukemia inhibitory factor (LIF) and ciliary neurotrophic factor (CNTF) decrease the number of primary dendrites and total dendritic arbor, without affecting axonal growth and neuronal survival. In addition, these cytokines can lead to retraction of pre-existing dendrites through the activation of STAT proteins [145–147]. In addition to activating STATs, IFNg activates Rit, (a small GTPase related to Ras GTPase) and p38-MAPK pathway to effect the dendritic retraction observed in these neurons [148]. Rit is expressed in sympathetic neurons and has opposite effects on axonal and dendritic growth. Dominant negative Rit transgenes decrease axonal elongation but enhance BMP-7 induced dendritic growth in an ERK-signaling dependent manner and constitutively active Rit enhances number of axons and axonal branching in sympathetic neurons while inhibiting dendritic growth [149].

3.5 Cytoskeletal proteins

Dendritic growth and remodeling requires changes to the actin and microtubule cytoskeleton [150, 151]. Signaling pathways downstream of Rho GTPases act as intermediates to connect extracellular signals and actin cytoskeletal remodeling during dendritic growth [152]. In cultured sympathetic neurons, BMP-7 treatment increases the GTP bound RhoA [153] and decreases GTP-bound Rit [149], with no effects on other small GTPases. In cultured SCG neurons, BMP-7 induced dendritic growth requires the activation of RhoA [153], suggesting that activation of this GTPase may be the link to actin cytoskeleton remodeling necessary for dendritic growth.

The microtubule polarity in axons is different from that in dendrites. Unlike microtubules in axons, which have a uniform polarity, microtubules in dendrites have a mixed orientation that is driven by the different motor proteins [154, 155]. A kinesin related motor protein kinesin 6 (also known as CHO/MKLP1) mRNA and protein are expressed in cultured embryonic sympathetic neuron, with CHO/MKLP1 protein extending from the cell body to the newly formed dendrites [156, 157]. Two other kinesin related motors – Kinesin 5 (also known as Eg5 or Kif11) and kinesin 12 (also known as Kif15) – are also expressed in embryonic sympathetic neurons [158, 159], with kinesin 5 associating only with the microtubule cytoskeleton and kinesin 12 being enriched in the dendrites and associating with both actin and microtubule cytoskeleton [158–160]. Treatment with antisense oligonucleotides against kinesin 6 lead to an increase in axonal length but a decrease in dendritic width and inhibition of BMP-7- induced dendritic growth in these neurons [156, 157]. Knockdown of kinesin 12 in cultured embryonic SCG neurons using an siRNA lead to longer axons that are less branched than control neurons and decrease in dendritic width [157, 160]. Both kinesin 6 and 12 appear to be important for the mixed polarity of microtubules in the dendrites with a decrease in these kinesins leading t0 fewer minus-end directed microtubules in the dendrites and increased frequency of microtubule transport. Similar to the others kinesins, inhibition of kinesin 5 leads to increase in axonal length, however a decrease in kinesin5 also leads to axons being non-responsive to navigational cues [161, 162]. In addition to a decrease in dendritic width like other kinesins, a reduction in kinesin 5 causes a decrease in dendritic length, a small decrease in number of dendrites and a significant effect on dendritic morphology especially during dendritic maturation stages [163]. In contrast to other kinesin mutants, a decrease in kinesin 5 leads to more minus-end microtubules in the dendrites [163]. Interestingly, kinesin5 appears to be regulated by phosphorylation with more phosphorylated kinesin5 being localized to the dendrites, suggesting that kinesin5 could be a potential link between signaling pathways and the cytoskeletal remodeling during dendritogenesis [163].

3.6 Other signaling pathways involved in dendritic growth

Retinoic acid synthesis enzymes and signaling pathway components are expressed in embryonic sympathetic neurons and activation of retinoic acid signaling in embryonic SCG neurons *in vitro* inhibits BMP-7 induced dendritic growth [164]. Similarly, pituitary adenylate cyclase 38(PACAP 38) and vasoactive intestinal peptide (VIP) are released by the preganglionic neurons and in cultured perinatal sympathetic neurons, PACAP38 and VIP decrease the number of dendrites and the total dendritic arbor of BMP-7 treated neurons. This effect is mediated through activation of the PAC1 receptor leading to the phosphorylation and nuclear translocation of cyclic AMP response element binding (CREB) protein, with inhibition of adenylate cyclase activity leading to enhanced dendritic growth [165].

4. Limitations of current research and the path forward

Sympathetic neurons have been long regarded as an important model system for studying neuronal differentiation. Due to increased recognition of the importance of sympathetic nervous system dysregulation in many diseases, there has been a renewed interest in understanding the mechanisms controlling neuronal differentiation, target innervation and neuronal survival in these neurons. Significant strides have been made in understanding axonal growth over the past 70 years *in vitro and in vivo* but there are still questions that need further exploration. While we understand the importance of individual growth factors for axonal growth, many of the null mutants show some innervation of target tissues. That still leaves the question of how much each of these growth factors contribute to initiation and elongation of axons during normal development and how their signaling pathways are coordinated to regulate final axonal growth in different paravertebral and prevertebral ganglia.

In comparison to axonal growth, our understanding of dendritic growth in these neurons is much more limited. Most of the studies on sympathetic neurons have been limited to cultured SCG neurons, which leaves the question of whether similar signals are important for regulation of dendritic growth in other paravertebral and prevertebral ganglia. Even in the SCG, many disparate signaling pathways including BMP, NGF, cytokine, ROS, ubiquitin-proteasome, etc. have been shown to control the dendritic tree *in vitro*. However, it is unclear which of these interactions are crucial for dendritic arborization *in vivo* in the SCG and how these pathways coordinately regulate dendritogenesis.

Finally, additional whole genome analysis looking at transcripts, proteins and non-coding RNAs is needed to fully understand the downstream mediators of both axogenesis and dendritogenesis to identify the common regulators controlling neuronal polarity and function in these neurons.

Acknowledgements

This work was supported through Saint Mary's College internal research funding through Faculty Development Fund and Summer Research Fund.

Author details

Vidya Chandrasekaran
Saint Mary's College of California, Moraga, CA, USA

*Address all correspondence to: vc5@stmarys-ca.edu

IntechOpen

© 2022 The Author(s). Licensee IntechOpen. This chapter is distributed under the terms of the Creative Commons Attribution License (http://creativecommons.org/licenses/by/3.0), which permits unrestricted use, distribution, and reproduction in any medium, provided the original work is properly cited.

References

[1] Goldstein DS. Differential responses of components of the autonomic nervous system. Handbook of Clinical Neurology. 2013;**117**:13-22

[2] Espinosa-Medina I, Saha O, Boismoreau F, et al. The sacral autonomic outflow is sympathetic. Science. 2016; **354**:893-897

[3] Merola A, Romagnolo A, Rosso M, et al. Autonomic dysfunction in Parkinson's disease: A prospective cohort study. Movement Disorders. 2018;**33**:391-397

[4] Goldstein DS, Robertson D, Esler M, Straus SE, Eisenhofer G. Dysautonomias: Clinical disorders of the autonomic nervous system. Annals of Internal Medicine. 2002;**137**:753

[5] Chu CC, Tranel D, Damasio AR, et al. The autonomic-related cortex: pathology in Alzheimer's disease. Cereb Cortex. 1997;7:86-95

[6] Jensen-Dahm C, Waldemar G, Staehelin Jensen T, et al. Autonomic dysfunction in patients with mild to moderate Alzheimer's disease. Journal of Alzheimer's Disease. 2015;**47**:681-689

[7] Vinik AI, Maser RE, Mitchell BD, et al. Diabetic autonomic neuropathy. Diabetes Care. 2003;**26**:1553-1579

[8] Adamec I, Habek M. Autonomic dysfunction in multiple sclerosis. Clinical Neurology and Neurosurgery. 2013;**115**:S73-78. DOI: 10.1016/j.clineuro.2013.09.026

[9] Racosta JM, Kimpinski K. Autonomic dysfunction, immune regulation, and multiple sclerosis. Clinical Autonomic Research. 2016;**26**:23-31

[10] Goldberger JJ, Arora R, Buckley U, et al. Autonomic nervous system dysfunction: JACC focus seminar. Journal of the American College of Cardiology. 2019;**73**:1189

[11] Rafanelli M, Walsh K, Hamdan MH, et al. Autonomic dysfunction: Diagnosis and management. Handbook of Clinical Neurology. 2019;**167**:123-137

[12] Chan WH, Anderson CR, Gonsalvez DG. From proliferation to target innervation: signaling molecules that direct sympathetic nervous system development. Cell and Tissue Research. 2018;**372**:171-193. DOI: 10.1007/s00441-017-2693-x

[13] Rubin E. Development of the rat superior cervical ganglion: Ganglion cell maturation. The Journal of Neuroscience. 1985;**5**:673-684

[14] Rubin E. Development of the rat superior cervical ganglion: Ingrowth of preganglionic axons. The Journal of Neuroscience. 1985;**5**:685

[15] Rubin E. Development of the rat superior cervical ganglion: Initial stages of synapse formation. The Journal of Neuroscience. 1985;**5**:697-704

[16] Yang XM, Toma JG, Bamji SX, et al. Autocrine hepatocyte growth factor provides a local mechanism for promoting axonal growth. The Journal of Neuroscience. 1998;**18**:8369-8381

[17] Maina F, Hilton MC, Andres R, et al. Multiple roles for hepatocyte growth factor in sympathetic neuron development. Neuron. 1998;**20**:835-846

[18] Maina F, Klein R. Hepatocyte growth factor, a versatile signal for developing

neurons. Nature Neuroscience. 1999; **2**:213-217

[19] Maina F, Panté G, Helmbacher F, et al. Coupling met to specific pathways results in distinct developmental outcomes. Molecular Cell. 2001; **7**:1293-1306

[20] Thompson J, Dolcet X, Hilton M, et al. HGF promotes survival and growth of maturing sympathetic neurons by PI-3 kinase- and MAP kinase-dependent mechanisms. Molecular and Cellular Neurosciences. 2004;**27**:441-452

[21] Baloh RH, Enomoto H, Johnson EM, et al. The GDNF family ligands and receptors—Implications for neural development. Current Opinion in Neurobiology. 2000;**10**:103-110

[22] Airaksinen MS, Saarma M. The GDNF family: Signalling, biological functions and therapeutic value. Nature Reviews Neuroscience. 2002;**3**:383-394

[23] Miwa K, Lee JK, Takagishi Y, et al. Axon guidance of sympathetic neurons to cardiomyocytes by glial cell line-derived neurotrophic factor (GDNF). PLoS One. 2013;**8**:e65202

[24] Ernsberger U. The role of GDNF family ligand signalling in the differentiation of sympathetic and dorsal root ganglion neurons. Cell and Tissue Research. 2008;**333**:353

[25] Baloh RH, Tansey MG, Lampe PA, et al. Neurotrophic factors represent a heterogeneous group of Sanicola et al. Neuron. 1998;**21**:1291-1302

[26] Enomoto H, Crawford PA, Gorodinsky A, et al. RET signaling is essential for migration, axonal growth and axon guidance of developing sympathetic neurons. Development. 2001;**128**:3963-3974

[27] Forgie A, Doxakis E, Buj-Bello A, et al. Differences and developmental changes in the responsiveness of PNS neurons to GDNF and neurturin. Molecular and Cellular Neurosciences. 1999;**13**:430-440

[28] Honma Y, Araki T, Gianino S, et al. Artemin is a vascular-derived neurotropic factor for developing sympathetic neurons. Neuron. 2002;**35**:267-282

[29] Nishino J, Mochida K, Ohfuji Y, et al. GFR alpha3, a component of the artemin receptor, is required for migration and survival of the superior cervical ganglion. Neuron. 1999;**23**:725-736

[30] Yan H, Newgreen DF, Young HM. Developmental changes in neurite outgrowth responses of dorsal root and sympathetic ganglia to GDNF, neurturin, and artemin. Developmental Dynamics. 2003;**227**:395-401

[31] Andres R, Forgie A, Wyatt S, et al. Multiple effects of artemin on sympathetic neurone generation, survival and growth. Development. 2001;**128**:3685-3695

[32] Durbec P, Marcos-Gutierrez CV, Kilkenny C, et al. GDNF signalling through the Ret receptor tyrosine kinase. Nature. 1996;**381**:789-793

[33] Reichardt LF. Neurotrophin-regulated signalling pathways. Philosophical Transactions of the Royal Society B. 2006;**361**:1545

[34] Skaper SD. Neurotrophic factors: An overview. Methods in Molecular Biology. 2018;**1727**:1-17

[35] Huang EJ, Reichardt LF. Neurotrophins: Roles in neuronal development and function. Annual Review of Neuroscience. 2001;**24**:677

[36] Davies AM. Neurotrophins giveth and they taketh away. Nature Neuroscience. 2008;**11**:627-628

[37] Davies AM. Extracellular signals regulating sympathetic neuron survival and target innervation during development. Autonomic Neuroscience. 2009;**151**:39-45

[38] Lee R, Kermani P, Teng KK, et al. Regulation of cell survival by secreted proneurotrophins. Science. 2001;**294**:1945-1948

[39] Korsching S, Thoenen H. Nerve growth factor in sympathetic ganglia and corresponding target organs of the rat: Correlation with density of sympathetic innervation. Proceedings of the National Academy of Sciences of the United States of America. 1983;**80**:3513

[40] Korsching S, Thoenen H. Developmental changes of nerve growth factor levels in sympathetic ganglia and their target organs. Developmental Biology. 1988;**126**:40-46

[41] Belliveau DJ, Krivko I, Kohn J, et al. NGF and neurotrophin-3 both activate TrkA on sympathetic neurons but differentially regulate survival and neuritogenesis. The Journal of Cell Biology. 1997;**136**:375-388

[42] Levi-Montalcini R. The nerve growth factor 35 years later. Science. 1987;**237**:1154-1162

[43] Shooter EM. Early days of the nerve growth factor proteins. Annual Review of Neuroscience. 2001;**24**:601-629

[44] Birren SJ, Lo L, Anderson DJ. Sympathetic neuroblasts undergo a developmental switch in trophic dependence. Development. 1993; **119**:597-610

[45] Glebova NO, Ginty DD. Growth and survival signals controlling sympathetic nervous system development. Annual Review of Neuroscience. 2005;**28**:191-222

[46] Unsicker K, Wiegandt H. Promotion of survival and neurite outgrowth of cultured peripheral neurons by exogenous lipids and detergents. Experimental Cell Research. 1988; **178**:377-389

[47] Edwards RH, Rutter WJ, Hanahan D. Directed expression of NGF to pancreatic beta cells in transgenic mice leads to selective hyperinnervation of the islets. Cell. 1989;**58**:161-170

[48] Hoyle GW, Mercer EH, Palmiter RD, et al. Expression of NGF in sympathetic neurons leads to excessive axon outgrowth from ganglia but decreased terminal innervation within tissues. Neuron. 1993;**10**:1019-1034

[49] Eldredge LC, Gao XM, Quach DH, et al. Abnormal sympathetic nervous system development and physiological dysautonomia in Egr3-deficient mice. Development. 2008;**135**:2949-2957

[50] Campenot RB. Local control of neurite development by nerve growth factor. Proceedings of the National Academy of Sciences of the United States of America. 1977;**74**:4516-4519

[51] Campenot RB. Development of sympathetic neurons in compartmentalized cultures. II. Local control of neurite survival by nerve growth factor. Developmental Biology. 1982;**93**:13-21

[52] Smeyne RJ, Klein R, Schnapp A, et al. Severe sensory and sympathetic neuropathies in mice carrying a disrupted Trk/NGF receptor gene. Nature. 1994;**368**:246-249

[53] Fagan AM, Zhang H, Landis S, et al. TrkA, but not TrkC, receptors are essential for survival of sympathetic neurons in vivo. The Journal of Neuroscience. 1996;**16**:6208-6218

[54] Crowley C, Spencer SD, Nishimura MC, et al. Mice lacking nerve growth factor display perinatal loss of sensory and sympathetic neurons yet develop basal forebrain cholinergic neurons. Cell. 1994;**76**:1001-1011

[55] Howard L, Wyatt S, Nagappan G, et al. ProNGF promotes neurite growth from a subset of NGF-dependent neurons by a p75NTR-dependent mechanism. Development. 2013;**140**:2108-2117

[56] Knudson CM, Tung KSK, Tourtellotte WG, et al. Bax-deficient mice with lymphoid hyperplasia and male germ cell death. Science. 1995;**270**:96-99

[57] Deckwerth TL, Elliott JL, Knudson CM, et al. BAX is required for neuronal death after trophic factor deprivation and during development. Neuron. 1996;**17**:401-411

[58] Glebova NO, Ginty DD. Heterogeneous requirement of NGF for sympathetic target innervation in vivo. The Journal of Neuroscience. 2004;**24**:743

[59] Tsui-Pierchala BA, Ginty DD. Characterization of an NGF-P-TrkA retrograde-signaling complex and age-dependent regulation of TrkA phosphorylation in sympathetic neurons. The Journal of Neuroscience. 1999;**19**:8207-8218

[60] Kuruvilla R, Zweifel LS, Glebova NO, et al. A neurotrophin signaling cascade coordinates sympathetic neuron development through differential control of TrkA trafficking and retrograde signaling. Cell. 2004;**118**:243-255

[61] Harrington AW, St. Hillaire C, Zweifel LS, et al. Recruitment of actin modifiers to TrkA endosomes governs retrograde NGF signaling and survival. Cell. 2011;**146**:421

[62] Ginty DD, Segal RA. Retrograde neurotrophin signaling: Trk-ing along the axon. Current Opinion in Neurobiology. 2002;**12**:268-274

[63] Riccio A, Pierchala BA, Ciarallo CL, et al. An NGF-TrkA-mediated retrograde signal to transcription factor CREB in sympathetic neurons. Science. 1997;**277**:1097-1100

[64] Cosker KE, Courchesne SL, Segal RA. Action in the axon: Generation and transport of signaling endosomes. Current Opinion in Neurobiology. 2008;**18**:270

[65] MacInnis BL, Senger DL, Campenot RB. Spatial requirements for TrkA kinase activity in the support of neuronal survival and axon growth in rat sympathetic neurons. Neuropharmacology. 2003;**45**:995-1010

[66] Campenot RB. NGF uptake and retrograde signaling mechanisms in sympathetic neurons in compartmented cultures. Results and Problems in Cell Differentiation. 2009;**48**:141-158

[67] Howe CL, Mobley WC. Long-distance retrograde neurotrophic signaling. Current Opinion in Neurobiology. 2005;**15**:40-48

[68] Campenot RB. Local control of neurite sprouting in cultured sympathetic neurons by nerve growth factor. Developmental Brain Research. 1987;**37**:293-301

[69] Atwal JK, Massie B, Miller FD, et al. The TrkB-Shc site signals neuronal survival and local axon growth via MEK and P13-kinase. Neuron. 2000;**27**:265-277

[70] Goold RG, Gordon-Weeks PR. NGF activates the phosphorylation of MAP1B by GSK3β through the TrkA receptor and not the p75NTR receptor. Journal of Neurochemistry. 2003;**87**:935-946

[71] Chen B, Hammonds-Odie L, Perron J, et al. SHP-2 mediates target-regulated axonal termination and NGF-dependent Neurite growth in sympathetic neurons. Developmental Biology. 2002;**252**:170

[72] Bodmer D, Levine-Wilkinson S, Richmond A, et al. Wnt5a mediates nerve growth factor-dependent axonal branching and growth in developing sympathetic neurons. The Journal of Neuroscience. 2009;**29**:7569-7581

[73] Li L, Eldredge LC, Quach DH, et al. Egr3 dependent sympathetic target tissue innervation in the absence of neuron death. PLoS One. 2011;**6**:e2569. DOI: 10.1371/journal.pone.0025696

[74] Quach DH, Oliveira-Fernandes M, Gruner KA, et al. A sympathetic neuron autonomous role for Egr3-mediated gene regulation in dendrite morphogenesis and target tissue innervation. The Journal of Neuroscience. 2013;**33**:4570

[75] Crerar H, Scott-Solomon E, Bodkin-Clarke C, et al. Regulation of NGF signaling by an axonal Untranslated mRNA. Neuron. 2019;**102**:553-563

[76] Scott-Solomon E, Kuruvilla R. Prenylation of axonally translated Rac1 controls NGF-dependent axon growth. Developmental Cell. 2020;**53**:691

[77] Suo D, Park J, Harrington AW, et al. Coronin-1 is a neurotrophin endosomal effector that is required for developmental competition for survival. Nature Neuroscience. 2014;**17**:36-45. DOI: 10.1038/nn.3593

[78] Suo X, Park J, Young S, et al. Coronin-1 and calcium signaling governs sympathetic final target innervation. The Journal of Neuroscience. 2015;**35**:3893-3902

[79] Schecterson LC, Bothwell M. Novel roles for neurotrophins are suggested by BDNF and NT-3 mRNA expression in developing neurons. Neuron. 1992;**9**:449-463

[80] Dechant G, Barde Y-A. The neurotrophin receptor p75NTR: Novel functions and implications for diseases of the nervous system. Nature Neuroscience. 2002;**5**:1131-1136

[81] Dechant G, Rodriguez-Tebar A, Kolbeck R, et al. Specific high-affinity receptors for neurotrophin-3 on sympathetic neurons. The Journal of Neuroscience. 1993;**13**:2610-2616

[82] Ernfors P, Lee KF, Kucera J, et al. Lack of neurotrophin-3 leads to deficiencies in the peripheral nervous system and loss of limb proprioceptive afferents. Cell. 1994;**77**:503-512

[83] Fariñas I, Jones KR, Backus C, et al. Severe sensory and sympathetic deficits in mice lacking neurotrophin-3. Nature. 1994;**369**:658-661

[84] Story GM, Dicarlo SE, Rodenbaugh DW, et al. Inactivation of one copy of the mouse neurotrophin-3 gene induces cardiac sympathetic deficits. Physiological Genomics. 2000;**2**(3):129-136

[85] Cui X, Jing J, Wu R, et al. Adipose tissue-derived neurotrophic factor 3 regulates sympathetic innervation and

thermogenesis in adipose tissue. Nature Communications. 2021;**12**:5362. DOI: 10.1038/s41467-021-25766-2

[86] Keeler AB, Suo D, Park J, et al. Delineating neurotrophin-3 dependent signaling pathways underlying sympathetic axon growth along intermediate targets. Molecular and Cellular Neurosciences. 2017;**82**:66

[87] Kohn J, Aloyz RS, Toma JG, et al. Functionally antagonistic interactions between the TrkA and p75 neurotrophin receptors regulate sympathetic neuron growth and target innervation. The Journal of Neuroscience. 1999;**19**:5393-5408

[88] Causing CG, Gloster A, Aloyz R, et al. Synaptic innervation density is regulated by neuron-derived BDNF. Neuron. 1997;**18**:257-267

[89] Ernfors P, Kucera J, Lee KF, et al. Studies on the physiological role of brain-derived neurotrophic factor and neurotrophin-3 in knockout mice. The International Journal of Developmental Biology. 2003;**39**:799-807

[90] Singh KK, Park KJ, Hong EJ, et al. Developmental axon pruning mediated by BDNF-p75NTR-dependent axon degeneration. Nature Neuroscience. 2008;**11**:649-658. DOI: 10.1038/nn.2114

[91] Grivennikov SI, Kuprash DV, Liu ZG, et al. Intracellular signals and events activated by cytokines of the tumor necrosis factor superfamily: From simple paradigms to complex mechanisms. International Review of Cytology. 2006;**252**:129-161

[92] Hehlgans T, Pfeffer K. The intriguing biology of the tumour necrosis factor/tumour necrosis factor receptor superfamily: Players, rules and the games. Immunology. 2005;**115**:1

[93] Sun M, Fink PJ. A new class of reverse signaling costimulators belongs to the TNF family. Journal of Immunology. 2007;**179**:4307-4312

[94] Kisiswa L, Osório C, Erice C, et al. TNFα reverse signaling promotes sympathetic axon growth and target innervation. Nature Neuroscience. 2013;**16**:865

[95] Gutierrez H, O'Keeffe GW, Gavaldà N, et al. Nuclear factor κB signaling either stimulates or inhibits Neurite growth depending on the phosphorylation status of p65/RelA. The Journal of Neuroscience. 2008;**28**:8246

[96] Kisiswa L, Erice C, Ferron L, et al. T-type Ca2+ channels are required for enhanced sympathetic axon growth by TNFα reverse signalling. Open Biology. 2017;**7**:160288. DOI: 10.1098/RSOB.160288

[97] Gutierrez H, Kisiswa L, O'Keeffe GW, et al. Regulation of neurite growth by tumour necrosis superfamily member RANKL. Open Biology. 2013;**3**:120150. DOI: 10.1098/RSOB.120150

[98] O'Keeffe GW, Gutierrez H, Pandolfi PP, et al. NGF-promoted axon growth and target innervation requires GITRL-GITR signaling. Nature Neuroscience. 2008;**11**:135-142

[99] McKelvey L, Gutierrez H, Nocentini G, et al. The intracellular portion of GITR enhances NGF-promoted neurite growth through an inverse modulation of Erk and NF-κB signalling. Biology Open. 2012;**1**:1016-1023

[100] Howard L, Wosnitzka E, Okakpu D, et al. TWE-PRIL reverse signalling suppresses sympathetic axon growth

and tissue innervation. Development. 2018;**145**:dev165936. DOI: 10.1242/DEV.165936

[101] Calhan OY, Wyatt S, Davies AM. CD40L reverse signaling suppresses prevertebral sympathetic axon growth and tissue innervation. Developmental Neurobiology. 2019;**79**:949-962

[102] McWilliams TG, Howard L, Wyatt S, et al. Regulation of autocrine signaling in subsets of sympathetic NeuronsHas regional effects on tissue innervation. Cell Reports. 2015;**10**:1443

[103] Erice C, Calhan OY, Kisiswa L, et al. Regional differences in the contributions of TNF reverse and forward signaling to the establishment of sympathetic innervation. Developmental Neurobiology. 2019;**79**:317

[104] Lein PJ, Higgins D. Laminin and a basement membrane extract have different effects on axonal and dendritic outgrowth from embryonic rat sympathetic neurons in vitro. Developmental Biology. 1989;**136**:330-345

[105] Lein PJ, Higgins D, Turner DC, et al. The NC1 domain of type IV collagen promotes axonal growth in sympathetic neurons through interaction with the alpha 1 beta 1 integrin. The Journal of Cell Biology. 1991;**113**:417-428

[106] Osterhout DJ, Frazier WA, Higgins D. Thrombospondin promotes process outgrowth in neurons from the peripheral and central nervous systems. Developmental Biology. 1992;**150**:256-265

[107] DeFreitas MF, Yoshida CK, Frazier WA, et al. Identification of integrin alpha 3 beta 1 as a neuronal thrombospondin receptor mediating neurite outgrowth. Neuron. 1995;**15**:333-343

[108] Lein P, Guo X, Hedges AM, et al. The effects of extracellular matrix and osteogenic Protein-1 on the morphological differentiation of rat sympathetic neurons. International Journal of Developmental Neuroscience. 1996;**14**:203-215

[109] Tang D, Goldberg DJ. Bundling of microtubules in the growth cone induced by laminin. Molecular and Cellular Neurosciences. 2000;**15**:303-313

[110] Gardner RT, Habecker BA. Infarct-derived chondroitin sulfate proteoglycans prevent sympathetic Reinnervation after cardiac ischemia-reperfusion injury. The Journal of Neuroscience. 2013;**33**:7175

[111] Bai Y, Hart RP. Cultured sympathetic neurons express functional interleukin-1 receptors. Journal of Neuroimmunology. 1998;**91**:43-54

[112] Nolan AM, Nolan YM, O'Keeffe GW. IL-1β inhibits axonal growth of developing sympathetic neurons. Molecular and Cellular Neurosciences. 2011;**48**:142-150

[113] De Chaves EP, Bussiere M, MacInnis B, et al. Ceramide inhibits axonal growth and nerve growth factor uptake without compromising the viability of sympathetic neurons. The Journal of Biological Chemistry. 2001;**276**:36207-36214

[114] Posse de Chaves EI, Bussière M, Vance DE, et al. Elevation of ceramide within distal neurites inhibits neurite growth in cultured rat sympathetic neurons. Journal of Biological Chemistry. 1997;**272**:3028-3035. DOI: 10.1074/jbc.272.5.3028

[115] Voyvodic JT. Development and regulation of dendrites in the rat superior cervical ganglion. The Journal of Neuroscience. 1987;7:904-912

[116] Snider WD. Nerve growth factor enhances dendritic arborization of sympathetic ganglion cells in developing mammals. The Journal of Neuroscience. 1988;**8**:2628-2634

[117] Purves D, Hume RI. The relation of postsynaptic geometry to the number of presynaptic axons that innervate autonomic ganglion. The Journal of Neuroscience. 1981;**1**:441-452

[118] Purves D, Hadley RD, Voyvodic JT. Dynamic changes in the dendritic geometry of individual neurons visualized over periods of up to three months in the superior cervical ganglion of living mice. The Journal of Neuroscience. 1986;**6**:1051-1060

[119] Ruit KG, Osborne PA, Schmidt RE, et al. Nerve growth factor regulates sympathetic ganglion cell morphology and survival in the adult mouse. The Journal of Neuroscience. 1990; **10**:2412-2419

[120] Chandrasekaran V, Lein PJ. Regulation of dendritogenesis in sympathetic neurons. In: Autonomic Nervous System. Rijeka: InTech; 2018. DOI: 10.5772/intechopen.80480

[121] Katagiri T, Watabe T. Bone Morphogenetic Proteins. Cold Spring Harb Perspect Biology. 2016;**8**:a021899. DOI: 10.1101/cshperspect.a021899

[122] Wotton D, Massague J. Transcriptional control by the TGF- b / Smad signaling system. The EMBO Journal. 2000;**19**:1745-1754

[123] Derynck R, Zhang YE. Smad-dependent and Smad-independent pathways in TGF-β family signalling. Nature. 2003;**425**:577-584

[124] Guo X, Rueger D, Higgins D. Osteogenic protein-1 and related bone morphogenetic proteins regulate dendritic growth and the expression of microtubule-associated protein-2 in rat sympathetic neurons. Neuroscience Letters. 1998;**245**:131-134

[125] Lein PJ, Beck HN, Chandrasekaran V, et al. Glia induce dendritic growth in cultured sympathetic neurons by modulating the balance between bone morphogenetic proteins (BMPs) and BMP antagonists. The Journal of Neuroscience. 2002; **22**:10377-10387

[126] O'Keeffe GW, Gutierrez H, Howard L, et al. Region-specific role of growth differentiation factor-5 in the establishment of sympathetic innervation. Neural Development. 2016;**11**:4

[127] Zhang D, Mehler MF, Song Q, et al. Development of bone morphogenetic protein receptors in the nervous system and possible roles in regulating trkC expression. The Journal of Neuroscience. 1998;**18**:3314-3326

[128] Beck HN, Drahushuk K, Jacoby DB, et al. Bone morphogenetic protein-5 (BMP-5) promotes dendritic growth in cultured sympathetic neurons. BMC Neuroscience. 2001;**2**:12. DOI: 10.1186/1471-2202-2-12

[129] Lein P, Johnson M, Guo X, et al. Osteogenic protein-1 induces dendritic growth in rat sympathetic neurons. Neuron. 1995;**15**:597-605

[130] Majdazari A, Stubbusch J, Müller CM, et al. Dendrite complexity of sympathetic neurons is controlled during postnatal development by BMP signaling. The Journal of Neuroscience. 2013;**33**:15132-15144

[131] Ten Dijke P, Hill CS. New insights into TGF-β–Smad signalling. Trends in Biochemical Sciences. 2004;**29**:265-273

[132] Garred MM, Wang MM, Guo X, et al. Transcriptional responses of cultured rat sympathetic neurons during BMP-7-induced dendritic growth. PLoS One. 2011;**6**:e21754

[133] Pravoverov K, Whiting K, Thapa S, et al. MicroRNAs are Necessary for BMP-7-induced Dendritic Growth in Cultured Rat Sympathetic Neurons. Cellular and Molecular Neurobiology. 2019;**39**:917-934. DOI: 10.1007/s10571-019-00688-2

[134] Courter LA, Shaffo FC, Ghogha A, et al. BMP7-induced dendritic growth in sympathetic neurons requires p75 NTR signaling. Developmental Neurobiology. 2016;**76**:1003-1013

[135] Guo X, Lin Y, Horbinski C, et al. Dendritic growth induced by BMP-7 requires Smad1 and proteasome activity. Journal of Neurobiology. 2001;**48**:120-130

[136] Chandrasekaran V, Lea C, Sosa JC, et al. Reactive oxygen species are involved in BMP-induced dendritic growth in cultured rat sympathetic neurons. Molecular and Cellular Neurosciences. 2015;**67**:116-125

[137] Vaillant AR, Zanassi P, Walsh GS, et al. Signaling mechanisms underlying reversible, activity-dependent dendrite formation. Neuron. 2002;**34**:985-998

[138] Dedhar S, Williams B, Hannigan G. Integrin-linked kinase (ILK): A regulator of integrin and growth-factor signalling. Trends in Cell Biology. 1999;**9**:319-323

[139] Naska S, Park KJ, Hannigan GE, et al. An essential role for the integrin-linked kinase-glycogen synthase kinase-3 beta pathway during dendrite initiation and growth. The Journal of Neuroscience. 2006;**26**:13344-13356

[140] Bruckenstein DA, Higgins D. Morphological differentiation of embryonic rat sympathetic neurons in tissue culture. Developmental Biology. 1988;**128**:337-348

[141] Stachowiak MK, Fang X, Myers JM, et al. Integrative nuclear FGFR1 signaling (INFS) as a part of a universal ?Feed-forward-and-gate? Signaling module that controls cell growth and differentiation. Journal of Cellular Biochemistry. 2003;**90**:662-691

[142] Horbinski C, Stachowiak EK, Chandrasekaran V, et al. Bone morphogenetic protein-7 stimulates initial dendritic growth in sympathetic neurons through an intracellular fibroblast growth factor signaling pathway. Journal of Neurochemistry. 2002;**80**:54-63. DOI: 10.1046/j.0022-3042.2001.00657.x

[143] Kim I-J, Drahushuk KM, Kim W-Y, et al. Extracellular signal-regulated kinases regulate dendritic growth in rat sympathetic neurons. The Journal of Neuroscience. 2004;**24**:3304-3312

[144] O'Shea JJ, Gadina M, Kanno Y. Cytokine signaling: Birth of a pathway. Journal of Immunology. 2011;**187**:5475-5478

[145] Guo X, Metzler-Northrup J, Lein P, et al. Leukemia inhibitory factor and ciliary neurotrophic factor regulate dendritic growth in cultures of rat sympathetic neurons. Brain Research Developmental Brain Research. 1997;**104**:101-110

[146] Guo X, Chandrasekaran V, Lein P, et al. Leukemia inhibitory factor and ciliary neurotrophic factor cause dendritic retraction in cultured rat sympathetic neurons. The Journal of Neuroscience. 1999;**19**:2113-2121

[147] Kim I-J, Beck HN, Lein PJ, et al. Interferon gamma induces retrograde

dendritic retraction and inhibits synapse formation. The Journal of Neuroscience. 2002;**22**:4530-4539

[148] Andres DA, Shi G-X, Bruun D, et al. Rit signaling contributes to interferon-γ-induced dendritic retraction via p38 mitogen-activated protein kinase activation. Journal of Neurochemistry. 2008;**107**:1436-1447

[149] Lein PJ, Guo X, Shi GX, et al. The novel GTPase Rit differentially regulates axonal and dendritic growth. The Journal of Neuroscience. 2007;**27**:4725-4736

[150] Higgins D, Burack M, Lein P, et al. Mechanisms of neuronal polarity. Current Opinion in Neurobiology. 1997;7:599-604

[151] Konietzny A, Bär J, Mikhaylova M. Dendritic actin cytoskeleton: Structure, functions, and regulations. Frontiers in Cellular Neuroscience. 2017;**11**:147

[152] Van Aelst L, Cline HT. Rho GTPases and activity-dependent dendrite development. Current Opinion in Neurobiology. 2004;**14**:297-304

[153] Kim W-Y, Gonsiorek EA, Barnhart C, et al. Statins decrease dendritic arborization in rat sympathetic neurons by blocking RhoA activation. Journal of Neurochemistry. 2009;**108**:1057-1071

[154] Rao AN, Baas PW. Polarity Sorting of Microtubules in the Axon. Trends in Neurosciences. 2018;**41**:77-88. DOI: 10.1016/j.tins.2017.11.002

[155] Baas PW. The role of motor proteins in establishing the microtubule arrays of axons and dendrites. Journal of Chemical Neuroanatomy. 1998;**14**:175-180

[156] Sharp DJ, Yu W, Ferhat L, et al. Identification of a microtubule-associated motor protein essential for dendritic differentiation. The Journal of Cell Biology. 1997;**138**:833

[157] Lin S, Liu M, Mozgova OI, et al. Mitotic motors Coregulate microtubule patterns in axons and dendrites. The Journal of Neuroscience. 2012;**32**:14033

[158] Ferhat L, Cook C, Chauviere M, et al. Expression of the mitotic motor protein Eg5 in Postmitotic neurons: Implications for neuronal development. The Journal of Neuroscience. 1998;**18**:7822

[159] Buster DW, Baird DH, Yu W, et al. Expression of the mitotic kinesin Kif15 in postmitotic neurons: Implications for neuronal migration and development. Journal of Neurocytology. 2003;**32**:79-96

[160] Liu M, Nadar VC, Kozielski F, et al. Kinesin-12, a mitotic microtubule-associated motor protein, impacts axonal growth, navigation, and branching. The Journal of Neuroscience. 2010;**30**:14896

[161] Myers KA, Baas PW. Kinesin-5 regulates the growth of the axon by acting as a brake on its microtubule array. The Journal of Cell Biology. 2007;**178**:1081

[162] Nadar VC, Ketschek A, Myers KA, et al. Kinesin-5 is essential for growth cone turning. Current Biology. 2008;**18**:1972

[163] Kahn OI, Sharma V, González-Billault C, et al. Effects of kinesin-5 inhibition on dendritic architecture and microtubule organization. Molecular Biology of the Cell. 2015;**26**:66-77

[164] Chandrasekaran V, Zhai Y, Wagner M, et al. Retinoic acid regulates

the morphological development of sympathetic neurons. Journal of Neurobiology. 2000;**42**:383-393

[165] Drahushuk K, Connell TD, Higgins D. Pituitary adenylate cyclase-activating polypeptide and vasoactive intestinal peptide inhibit dendritic growth in cultured sympathetic neurons. The Journal of Neuroscience. 2002;**22**:6560-6569

Chapter 10

The Autonomic Nervous System, Sex Differences, and Chronobiology under General Anesthesia in *In Vivo* Experiments Involving Rats

Pavol Svorc Jr and Pavol Svorc

Abstract

The aim was to evaluate the current state of the autonomic nervous system (ANS) activity under general anesthesia using heart rate variability (HRV) in dependence on the light-dark (LD) cycle in healthy, sexually mature, spontaneously breathing, zoletil-anesthetized (30 mg/kg) Wistar rats of both sexes after a 4-week adaptation to an LD cycle (12 h:12 h). The animals were divided into four experimental groups according to sex and light period (n = 20 each). RR interval duration, spectral power at very-low-frequency (VLF), low-frequency (LF) and high-frequency (HF), total spectral power of HRV, and the LF/HF ratio were analyzed. Sympathetic and baroreceptor activity was decreased, and parasympathetic activity was increased in both sexes and in both light periods. Regarding sex differences, HRV was significantly lower in females versus males in the light period. In the dark period, females exhibited higher HRV than males. Regarding LD differences, in females, HRV was lower in the light versus the dark period, unlike males, in which HRV was higher in the dark versus the light period of the rat regimen day. Sex differences in the activity of the ANS were apparent in rats, persisted under general anesthesia, and were dependent on the LD cycle.

Keywords: HRV, sex, general anesthesia, chronobiology, rat

1. Introduction

The role of the autonomic nervous system (ANS) and its organ-specific functions have, in large part, been elucidated. Analysis of heart rate variability (HRV) is a popular tool for the assessment of autonomic cardiac control. Small periodic fluctuations in heart rate are well known to physicians and scientific investigators. Because these fluctuations are caused by the varying activity of the ANS, an examination of HRV is needed to obtain information about the functional status of the ANS. Heart rate and changes in heart rate are sensitive indicators of ANS function; therefore, cardiovascular autonomic regulation is considered to be the most reliable indicator of ANS activity.

HRV describes the beat-to-beat variation in heart rate and is used to quantify the interplay between the sympathetic and parasympathetic divisions of the ANS. Although patterns of HRV demonstrate considerable promise for clarifying issues in clinical applications, the inappropriate quantification and interpretation of these patterns may obscure critical issues or relationships, and may impede—rather than foster—the development of clinical applications [1].

HRV analysis, which supports the evaluation of successive RR intervals using electrocardiographic (ECG) methods, has been a powerful tool in the assessment of autonomic cardiac control [2]. For example, in humans, reduced HRV is associated with an increased risk for ventricular arrhythmia and has been shown to be an independent prognostic factor for mortality in patients with cardiac disease(s) [3, 4]. On the other hand, some studies have demonstrated that analysis of HRV spectral performance in rats is an ineffective method for detecting heart-related autonomic control disorders in some experimental models of myocardial infarction or diabetic neuropathy [5–7].

1.1 Chronobiology of HRV

The ANS is an important control system that affects the function of many organs, and its activity is affected by various factors, including age [8], sex, and internal processes, such as circadian rhythm and hormonal fluctuations that slowly rise and fall over the course of 24 h. Circadian fluctuations in HRV parameters in rats were confirmed in a study by Hashimoto et al. [9], who reported that sympathetic nerve activity predominates in the dark phase. The ratio of low frequency (LF) to high frequency (HF) demonstrated a nocturnal pattern, and the value in the dark phase was significantly higher than in the light phase. In 2001, Hashimoto et al. [10] extended the monitoring of circadian rhythmicity in HRV to diabetic rats. Although diabetic autonomic neuropathy modifies circadian rhythms in HRV in diabetic WBN/Kob rats, in healthy nondiabetic Wistar rats, significant light–dark (LD) differences were detected in some of the monitored HRV parameters. In both age-different and pre-diabetic Wistar and diabetic WBN/Kob rats, no LD differences were found in the LF parameter of HRV; however, significant LD differences in the HF parameter were detected, except in older diabetic rats. Significant LD differences were found in the LF/HF ratio, but only in prediabetic Wistar rats.

In a telemetry study, Mamalyga [11] described fluctuations in ANS activity during a 24 h period, in which the control groups of male rats exhibited the greatest predominance of sympathetic activity between 12:00 h and 24:00 h. Similarly, in this time range, the LF parameter and LF/HF ratio exhibited higher values, and the HF parameter of HRV exhibited lower values. Analysis of multiday ECG recordings demonstrated the predominance of different mechanisms of heart rhythm regulation in experimental and control rats over a 24 h period. More severe dysfunction of neuroautonomical mechanisms of regulation in experimental rats was reflected in circadian dynamics. Further evidence supporting the existence of circadian rhythms in ANS activity was obtained in a study by Hsieh et al. [12], who monitored various physiological signals after implantation of sensors into the abdomen of rats and were recorded without interruption for >10 days. There was no difference in sleep/wakefulness patterns, physical activity, body weight, and autonomic functioning assessed according to HRV among control, sham, and experimental rats. Continuous recording further revealed circadian rhythms in HRV parameters, namely a 24 h cycle in RR intervals, the total power of HRV, and HF and LF powers of the RR spectrum. As such, we believe that this information may be useful in future biobehavioral studies.

In common practice, experiments are performed during "regular" working hours, even after the synchronization of rats to the LD cycle (12 h:12 h). Although this synchronization is often described in the methods section of these studies, the time of day when the experiments are performed is not reported. Therefore, it is assumed that the experiments are performed during the day (i.e., during the light) and, thus, on "sleeping" rats in their inactive period of the regimen day. However, the question is, what are the reactions of animals in their active period if there are fluctuations in the functions of individual systems in both sexes? Is there alternative reactivity of these systems, or is there a uniform reaction in both sexes? Therefore, if sex differences in the results of various experimental studies are documented, it is necessary to respect this fact. As such, future studies should decode these questions and try to include females in experiments whenever possible.

In the planning stages and design of *in vivo* experiments, researchers often encounter multiple problems, one of which can be the actual methodology. Established and proven methods are often used and are precisely focused on the type of experiment, whereas other factors that may affect and, consequently, lead to misinterpretation of the results are not taken into account.

1.2 Anesthesia

However, approaches based on ECG recordings of animals in an anesthetic state are not ideal nor valid for HRV analysis due to significant heart rate fluctuations associated with impaired autonomic modulation of the heart [13, 14]. In addition, anesthesia may contribute an important additional risk for animal mortality under some pathological conditions such as myocardial infarction and diabetes mellitus [15–17]. General anesthesia weakens autonomic function and baroreflex control. This side effect should be avoided as much as possible because it limits the ability of the subject to respond to physiological challenges during surgery [18]. Therefore, any research intervention that could affect aspects of the ANS and its impact(s) on internal organs should take into account the anesthetic used. Intravenous anesthetics may have different qualitative and quantitative effects on the peripheral ANS and, thus, may alter the activity of the sympathetic or parasympathetic divisions of the ANS.

1.3 Sex

In the vast majority of experimental studies, only male rats are used; however, there is also the other sex (i.e., female), in which differences may already exist in the very essence of the monitored functional system and exhibit a different response to interventions. At the same time, the study of sex differences is a driving force for development and, in many cases, the basis of health and medicine. However, there are opinions that the investigation of sex differences is ineffectual and does not merit extensive research [19].

Although there are several reasons why female animals are omitted, the primary rationale is simple—males and females are biologically different. Among other reasons, some scientists consider males to be representative of humans and differences from male norms are considered to be atypical or abnormal. Others attempt to "protect" females from the adverse effects of various interventions [20]. Still, others generalize findings from males and females, regardless of differences and, generally speaking, most scientists use male rats because they want to avoid accounting for hormonal cycles in females, which may reduce the homogeneity of the study population and affect the impact of experimental interventions [21]. When females are included in experiments, two problems arise—the sample size is effectively halved—the economic aspect; and the dispersion of results increases. One explanation for the increased variance is the simple fact that males and females are different and these differences increase the range of variability. However, if males and females are mixed, scientists may find a beneficial effect of a tested drug, for example, that lowers blood pressure, in both sexes [19]. On the other hand, on obtaining results from *in vivo* experiments in rats, errors in general interpretation may arise due to differences related to sex, and these discrepancies occur not only in behavioral studies [22–24]. Furthermore, there are also sex-dependent differences in drug metabolism and the action of liver enzymes [25], in the internal environment [26], in the activity of the ANS and cardiovascular system [27, 28], and most probably, in other functions as well.

As such, whether to acknowledge sex differences in *in vivo* experiments involving rats becomes a legitimate concern. Presently, however, there are relatively little data regarding sex differences in ANS activity or ANS activity during anesthesia. During resting conditions, male rats exhibit a significantly higher heart rate and lower HRV parameters than female rats. This occurs not only during the active but also during the inactive phase of the daily cycle of rats [27]. Further *in vivo* rat studies have confirmed results reported by Koresh et al. [27]—that there are significant sex differences in HRV and depend on the LD cycle, even under conditions of general anesthesia [28]. LD differences with nonsignificantly lower HRV were found in females in the light part compared to the dark part of the regimen day, in contrast to males, in which HRV was significantly higher in the light part of the day.

These data support the concept that sex-based variations should also be taken into account, given that females in human and animal studies exhibit different mechanisms of cardiovascular regulation [29]. Although these data suggest that if there are sex differences in individual cardiovascular parameters, they are predominantly regulated by the ANS. Logically, therefore, if sex differences exist in cardiovascular activities, sex differences in the circadian oscillations of individual divisions of the ANS must also exist in parallel.

The aim of the present study was not to downplay or critique the excellent and valid results of experimental *in vivo* studies involving rats but to raise awareness to the possibility of improving the design of the experiments themselves, not only by respecting sex differences, but also chronobiological principles. Accordingly, the primary goal of this investigation was to determine whether there are sex differences in ANS activity, measured according to HRV dependence on the LD cycle (a parallel to the circadian rhythm) in healthy, sexually mature, spontaneously breathing, zoletil-anesthetized rats.

2. Materials and methods

2.1 Ethics approval

The present study conformed to the Guide for the Care and Use of Laboratory Animals published by the United States National Institutes of Health (publication number 85–23, revised 1996). The study protocol was approved by the Ethics Committee of the Medical Faculty of Safarik University (Kosice, Slovak Republic; permission number 2/05 and permission number ŠVPS SR: Ro4234/15–221).

2.2 Animals

The experiments were performed using Wistar albino rats (weight, 340 ± 40 g, 3–4 months of age) acquired from a breeding and vendor company (VELAZ, Koleč, Czech Republic, certificate number 70029/2013-MZE-17214), with veterinary registration number CZ 21760118.

2.3 Adaptation

The animals were quarantined for 2 weeks in the Laboratory of Research Biomodels of the Medical Faculty of Safarik's University in Košice (official number SK UCH 08018) and adapted to an LD cycle (12 h light:12 h dark [intensity of constant artificial illumination during the light period, 80 Lux]); 40–60% humidity; cage temperature 24°C; two animals/plastic cage for 4 weeks. The rats were fed a standard pellet diet, with *ad libitum* access to food and water. Animal handling was performed by the professional staff of the animal facility.

2.4 Anesthesia

Anesthesia (zoletil, 30 mg/kg, Virbac, France) was administered in prescribed doses in the adaptation room by intraperitoneal injection based on the weight of the animal. After testing the effect of anesthesia (loss of uprighting reflexes, reaction to painful stimulus), the animals were transferred to the operating room, where they were fixed to an experimental table on which subcutaneous electrodes were used to record ECG and HRV. Again, the depth of anesthesia was assessed depending on whether the painful stimulus caused noticeable motor movements (minimal limb movement and muscle tension change) or cardiovascular responses such as changes in heart rate or onset of heart rhythm disorders.

2.5 Experimental groups

The effect of the light period on the monitored parameters was examined after adaptation to an LD cycle, with the light period from 06:00 h to 18:00 h. The effect of the dark period was monitored after adaptation to the inverse setting of the LD cycle (i.e., with the light period from 18:00 h to 06:00 h). The animals were randomly divided into four experimental groups (n = 20 each) according to sex and light conditions—group 1, female (light period); group 2, female (dark period); group 3, male (light period); and group 4, male (dark period). In *in vivo* experiments, at least 20 animals are valid sample size for statistical processing.

2.6 Protocol

HRV was analyzed using the ID Instruments computer system for biopotential recording from an average of 220 heart cycles, 20 minutes after administration of anesthesia at 09:00 h—12:00 h using separate animals. In analyzing HRV parameters, the focus was on the evaluation of RR interval duration spectral power at very-low-frequency (VLF, 0.003–0.04 Hz), low-frequency (LF, 0.04–0.15 Hz), and high-frequency (HF, 0.15–0.4 Hz), total spectral power of HRV, and the LF/HF ratio. The experiments were performed throughout the year and the results were averaged independently of the season and, in females, independently of the estral cycle. All animals (i.e., 20/20) were included in the statistical analysis. Before and after administration of the anesthetic, as well as during measurement, there were no adverse events or unexpected changes in HRV or ECG parameters, although considerable variability was observed. On completion of the measurement, the animals were transferred to the animal facility.

2.7 Statistical analysis

Data are expressed as mean ± standard deviation (SD). Data were analyzed using InStat (GraphPad, San Diego, CA, USA). The Tukey–Kramer test was used to compare data from the groups, and differences with $p < 0.05$ were considered to be statistically significant.

3. Results and discussion

3.1 RR interval

Evaluation of the RR interval can sometimes be problematic because different effect(s) of the anesthetic on this parameter has been described. The data reported in **Table 1** indicate values of the duration of the RR interval from telemetry studies and under different types of general anesthesia according to sex and dependence on the LD cycle (**Figure 1**).

Baseline RR interval analysis from telemetry studies [9, 7, 27, 30, 31] involving male Wistar rats, in which a chronobiological approach was applied, indicates that there is a circadian rhythm in the duration of RR intervals in rats, with a lower RR interval duration during the active (i.e., dark) period of the regimen day. Although adaptation of animals to the LD cycle was described in these articles, exactly what time of day the measurements were performed was not reported, nor whether they were average values from the entire 24 h period or only from certain time intervals the measurements were performed and recorded. The averaged results of baseline RR interval duration indicate that sex differences are exhibited in both the light and dark period of the rat regimen day; however, more experimental studies are needed to confirm this conclusion.

When comparing the duration of the RR interval in male rats from telemetry studies, it is clear that the shorter duration occurred during the dark period, which corresponds to a higher heart rate. Under zoletil anesthesia, a shorter RR interval was found in both light phases of the rat regimen day compared with values from telemetry studies, indicating a tachycardic effect of this anesthetic. The shortened duration of the RR interval corresponded to increased heart rate in both sexes in both lighted periods of the regimen day. Among females, LD differences were not observed in the duration of the RR interval (light, 142.30 ± 25.19 ms vs. dark, 134.97 ± 9.09 ms), in contrast to males, in which a significantly ($p < 0.001$) longer RR interval was recorded during the light part of the day (light, 145.05 ± 8.51 ms vs. dark, 124.68 ± 5.14 ms). Sex differences were found only in the dark (i.e., active) part of the day, with significantly shorter RR intervals in males. On the other hand, significant LD differences were maintained in males but eliminated in females. Addtitionally, compared to values reported in telemetry studies (**Table 1**), the RR interval was shorter, indicating a higher heart rate.

Our results, therefore, indicate that in zoletil-anesthetized rats, LD differences were maintained only in males but not in females. Considering the results of telemetry studies by Molcan et al. [50, 51], heart rate exhibits a significant circadian

	Light period		Dark period	
Anesthesia	**Female**	**Male**	**Female**	**Male**
Telemetry studies	168.7 (167.3–170.1) (n = 1)	163.2 (157–168.5) (n = 2)	140.2 (139.5–141) (n = 1)	145.9 (142.6–149.2) (n = 2)
Pentobarbital	177 (174–180) (n = 1)	—	165 (163–167) (n = 1)	—
Ketamine	271.1 (231.5–310.7) (n = 2)	—	213.1 (194.1–232.1) (n = 2)	—
Tribromoethanol	—	—	—	—
Thiopental	—	—	—	—
Urethane	—	—	—	—
Zoletil (Present study)	142.30 (117.1–167.5)	145.05 (136.5–153.6)	134.97 (125.9–144.1)	124.68 (119.5–129.8)

Data presented as the average RR interval duration (ms) (range); (n, number of experiments from which RR interval was evaluated).

Table 1.
Duration of RR interval from telemetry studies and for different types of general anesthesia according to sex and dependence on the light–dark cycle.

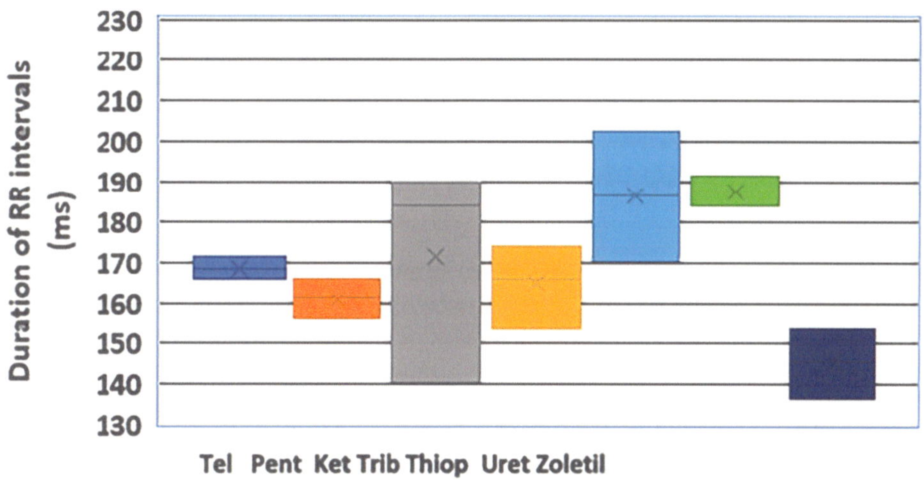

Figure 1.
Distribution of average values and ranges of RR intervals from telemetry studies and under different types of general anesthesia in male rats, without specification of synchronization to the light–dark cycle or the time of day when the experiments were performed. Tel – Telemetry studies (168.5(165.6–171.5), n = 3) [7, 30, 31]; pent – Pentobarbital (161.1(156.1–165.7), n = 6) [32–37]; Ket – Ketamine (183.8(140.2–189.8), n = 5) [38–42]; Trib – Tribromoethanol (166(154–174), n = 1) [43]; Thitop – Tiopenthal (186.2(170.3–202.1), n = 4) [44–47]; Uret – Urethane (187.5(183.7–191.2), n = 2) [48, 49]; zoletil – Zoletil anesthesia (145.05(136.5–153.6), n = 1) [presented results]. Data presented as average RR interval duration (ms) (range); (n, number of experiments from which RR interval was evaluated).

rhythm in non-anesthetized rats, in which the heart rate in the dark period fluctuated from 347 beats/min to 363 beats/min, and from 309 beats/min to 321 beats/min in the light period. Thus, it appears that although zoletil exerts a tachycardic effect, it can eliminate——or, at least modify——the circadian rhythm of heart rate, but only in females.

Such elimination or modification of LD differences in heart rate among females may also be partly explained by the greater sensitivity of females to acidosis, hypoxia, and hypercapnia under general anesthesia [52]. Previous studies have described the effect of hypoxia on the modulation of daily rhythmicity [53–57]. The fact that hypoxia modifies circadian oscillations of important variables, such as body temperature and metabolism, can lead to the expectation that the rhythms of many functions are interrupted by hypoxia on the basis of their relationship with the primary variables. Such a relationship likely contributes to a greater parasympathetic effect(s) on the heart [58]. Additionally, the effect of anesthetics can contribute to the loss or modification of rhythmicity. For example, in female rats under pentobarbital anesthesia, parasympathetic activity increases and sympathetic and baroreflex activity decreases; however, LD differences in heart rate are eliminated. Under ketamine/xylazine anesthesia, a preference toward parasympathetic activity was increased and sympathetic and baroreflex activity was depressed, resulting in significant bradycardia but with the maintenance of LD differences [59].

The paradox, under ketamine/xylazine anesthesia, therefore, remains—on the one hand, there is clearly evident increased parasympathetic activity and, on the other hand, increased heart rate. This paradox has been described by several authors [60–65], who assumed that stimulation of the vagal nerve releases catecholamines, which in turn can affect heart activity. This is also probably the case with zoletil anesthesia, which may have a similar effect on the release of catecholamines through higher parasympathetic activity, and is particularly evident in males in both light periods of the regimen day. Because sympathetic tone is significantly reduced and parasympathetic tone dominates, it is assumed that the duration of RR intervals is predominantly determined by the parasympathetic system.

3.2 HRV analysis

Despite the large variation in HRV spectral powers under zoletil anesthesia, in terms of sex differences, parasympathetic activity dominated in both sexes and in both light periods. In terms of sex differences, female HRV was significantly lower compared to males in the light period, while in the dark part of the regimen day, it was, in contrast, significantly higher in females compared to males (**Figure 2**).

Sympathetic activity dominates the normal life cycle of rats [7, 65] and zoletil anesthesia increases parasympathetic activity in both sexes. Similar results have been reported in previous studies. Administration of the anesthetic agent tribromoethanol in male Wistar rats [66], ketamine hydrochloride and diazepam in albino Wistar rats [67], and ketamine/xylazine and pentobarbital in females [59] resulted in predominant parasympathetic activity. However, our results indicate that precisely defining changes in HRV are difficult due to significant variability, which in turn makes it difficult to attribute sex differences. Thus, we agree with the opinion described in the introduction that approaches based on ECG recording under general anesthesia are not fully valid for HRV analysis.

In males under zoletil anesthesia—spectral power of HF (parasympathetic activity, $r = 0.96$) during the light and dark periods of the regimen day, spectral power of LF (baroreflex activity, $r = 0.95$), but also spectral power of HF (parasympathetic

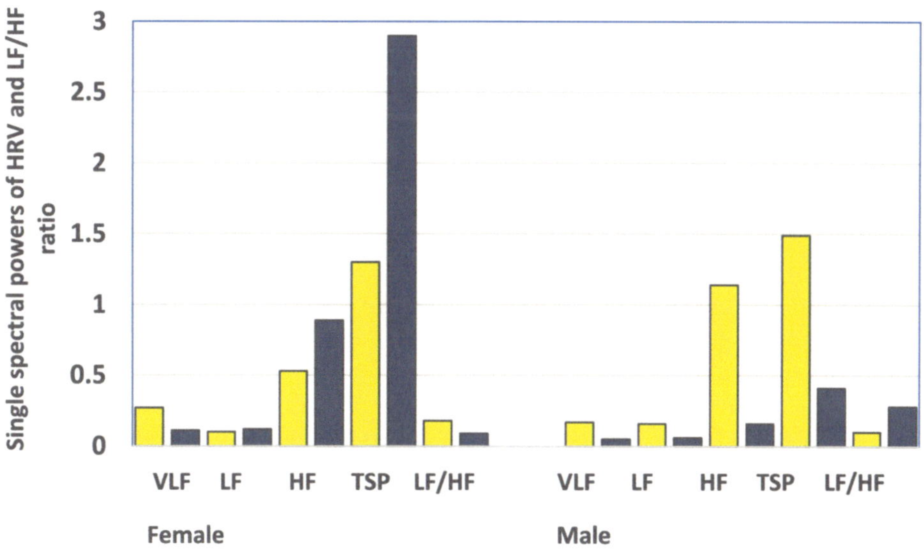

Figure 2.
Representation of heart rate variability (HRV) spectral powers in a rat model under zoletil anesthesia in both sexes. VLF – Spectral power of the very low frequency of HRV; LF - spectral power of the low frequency of HRV; HF - spectral power of high frequency of HRV; TSP – Total spectral power of HRV. Yellow columns – Light period of the rat regimen day; blue columns – Dark period of the rat regimen day.

activity, r = 0.81) significantly contributed to changes in the total spectral power of HRV. Sympathetic activity is practically not involved in the formation of the total spectral power of HRV. The participation of individual spectral powers, as well as the total spectral power of HRV in the duration of RR intervals, is minimal in both lighted periods of the rat regimen day (**Table 2**). After analysis of the dependence

Variable	Sex, light cycle			
	Female, light	Female, dark	Male, light	Male, dark
RR-VLF	**r = 0.54**	**r = 0.58**	r = 0.34	r = −0.13
RR-LF	**r = 0.47**	**r = 0.59**	r = 0.26	r = −0.12
RR-HF	r = 0.37	**r = 0.60**	r = 0.20	r = 0.06
RR-TSP	**r = 0.51**	**r = 0.61**	r = 0.28	r = 0.06
TSP-VLF	**r = 0.6**	**r = 0.87**	**r = 0.59**	r = 0.05
TSP-LF	**r = 0.99**	**r = 0.99**	**r = 0.59**	**r = 0.95**
TSP-HF	**r = 0.92**	**r = 0.89**	**r = 0.96**	**r = 0.81**

Bolded values indicate statistically significant dependence between single parameters. VLF – spectral power of the very low frequency of HRV; LF - spectral power of the low frequency of HRV; HF - spectral power of the high frequency of HRV; TSP – total spectral power of HRV.

Table 2.
Correlation coefficients of RR interval duration between spectral powers of heart rate variability (HRV) and the share of individual spectral powers in changes in the total spectral power of HRV.

of the duration of RR intervals on the total spectral power of HRV, we came to the conclusion that the duration of RR intervals (i.e., heart rate) is not regulated by the ANS in both light periods of the rat regimen day. We assume that other mechanisms are likely involved in the regulation of heart rate and are activated by zoletil.

In female rats under zoletil anesthesia, the spectral power of LF (baroreflex activity, r = 0.99) and spectral power of HF (parasympathetic activity, r = 0.92) contributed significantly to changes in the total spectral power of HRV during the light period of the day and during the dark period proportionally in all three spectral powers of HRV. Sympathetic activity in both lighted periods was involved in the formation of the total spectral power of HRV in females (**Table 2**). After analysis of the dependence of the duration of RR intervals on the total spectral power of HRV, we found that the duration of RR intervals (i.e., heart rate) was under the regulatory influence of the ANS in both lighted periods of the rat regimen day (light, r = 0.51; dark, r = 0.61) with proportional representation of all three spectral powers of HRV.

We conclude that there are sex differences in the total spectral power of HRV in zoletil-anesthetized Wistar rats. In the light period in females, HRV was significantly lower than in males, and vice versa in males in the dark period of the regimen day. This means that, in females, the myocardium may be more sensitive to ANS regulatory interventions in the dark versus the light period. It is generally accepted that decreased HRV is a predictor of myocardial infarction mortality and increased HRV is associated with decreased morbidity and mortality. From this point of view, in zoletil-anesthetized female Wistar rats, during the active (dark) period, there is greater electrical stability in the myocardium than during the inactive (light) period. On the contrary, in males, the heart more sensitive reacts to changes in ANS activity in the light versus the dark period of the regimen day.

In females, changes in HRV were the result of sympathetic (i.e., VLF) and baroreflex (i.e., LF) activities and, in males, parasympathetic (i.e., HF) activity dominated. Among females, changes in RR were primarily due to changes in HRV, whereas in males, changes in HRV had no effect on RR in both lighted parts of rat regimen day. The results of these studies show that not only sex—but also the time of day experiments are performed—also plays an important role [68]. However, supportive evidence of HRV changes in rats during a 24 h period is lacking.

The LF/HF ratio can be used to quantify the changing relationship between sympathetic and parasympathetic nerve activity (i.e., sympathetic-vagal balance) [69–71]. The exact interpretation of the LF/HF ratio also depends on the assumption that physiological interventions always cause mutual changes in parasympathetic and sympathetic activity.

Our results demonstrate that the LF/HF ratio depends on the light periods of the regimen day. In females in the light period, the LF/HF ratio was significantly higher and in the dark period, significantly lower than in males. These conclusions, however, should be interpreted with caution. In a study addressing the meaning of HRV examination, Billman [72] questioned the evaluation of the LF/HF ratio. The LF component of HRV does not provide a cardiac sympathetic response index, but rather reflects a complex and not a readily recognizable mixture of sympathetic, parasympathetic, and other unidentified factors with parasympathetic factors, which account for the largest part of the variability in this frequency range. As a result, it is difficult to recognize the physiological basis for LF/HF. In addition, a relatively large amount of data suggests that the spectral power of the HF component cannot be attributed solely to changes in cardiac vagal efferentation, further compromising the accurate interpretation of the LF/HF ratio [72].

4. Conclusions

In *in vivo* experiments, homeostatic regulatory mechanisms are not eliminated. This means that experimental results are a reflection of a direct but significantly intravariable response of the animals to the administration of anesthetic. On the evaluation of HRV in *in vivo* conditions, replacement and reduction of animals are not possible; however, knowledge about sex differences during anesthesia in the dependence on LD cycle in ANS activity may improve the quality of experimental design. There is little to no data regarding sex differences, and we do not have any data regarding changes in ANS activity depending on the LD cycle under general anesthesia. Further research is needed to assess the responses of other species because the effect of zoletil is essentially not described in experimental practices.

Based on our results, we conclude that under zoletil anesthesia, sympathetic (VLF) and baroreceptor (LF) activity were decreased, and parasympathetic (HF) activity was increased in both sexes and in both light periods. LD differences were preserved mainly in the HF component; thus, the circadian rhythm in parasympathetic activity likely also exists in both sexes. In terms of sex differences based on the total spectral power of HRV, our results suggest that HRV, in the light period of the rat regimen day, was significantly lower in females versus males. In the dark period, females exhibited higher HRV than males. In terms of LD differences, in females, HRV was lower in the light versus the dark period, unlike males, in which HRV was higher in the dark versus the light period of the rat regimen day.

Conflict of interest

The authors declare no conflict of interest.

Author details

Pavol Svorc Jr[1*] and Pavol Svorc[1,2]

1 Department of Physiology and Pathophysiology, Medical Faculty, Ostrava University, Ostrava, Czech Republic

2 Department of Physiology, Medical Faculty, Safarik's University, Kosice, Slovak Republic

*Address all correspondence to: pavol.svorc1@osu.cz

IntechOpen

© 2021 The Author(s). Licensee IntechOpen. This chapter is distributed under the terms of the Creative Commons Attribution License (http://creativecommons.org/licenses/by/3.0), which permits unrestricted use, distribution, and reproduction in any medium, provided the original work is properly cited.

References

[1] Singh B, Bharti N. Software tools for heart rate variability analysis. Int J Recent Sci Res. 2015;**6**:3501-3506. DOI: 10.24327/IJRSR

[2] Akselrod S, Gordon D, Ubel F, FA, Shannon DC, Berger AC, Cohen RJ. Power spectrum analysis of heart rate fluctuation: a quantitative probe of beat-to-beat cardiovascular control. Science. 1981;**213**:220-222. DOI: 10.1126/science.6166045

[3] Kleiger RE, Miller JP, Bigger JT Jr, Moss AJ. Decreased heartrate-variability and its association with increased mortality after acute myocardial infarction. The American Journal of Cardiology. 1987;**59**:256-262. DOI: 10.1016/0002-9149(87)90795-8

[4] Nolan J, Batin PD, Andrews R, Lindsay SJ, Brooksby P, Mullen M, et al. Prospective study of heart rate variability and mortality in chronic heart failure– Results of the United Kingdom heart failure evaluation and assessment of risk trial (UK-Heart). Circulation. 1998;**98**:1510-1516. DOI: 10.1161/01.cir.98.15.1510

[5] Krüger C, Landerer V, Zugck C, Ehmke H, Kübler W, Haass M. The bradycardic agent zatebradine enhances baroreflex sensitivity and heart rate variability in rats early after myocardial infarction. Cardiovascular Research. 2000;**45**:900-912. DOI: 10.1016/S0008-6363(99)00405-8

[6] Sanyal SN, Arita M, Ono K. Inhomogeneous derangement of cardiac autonomic nerve control in diabetic rats. Circulation Journal. 2002;**66**:283-288. DOI: 10.1253/circj.66.283

[7] Pereira-Junior PP, Marocolo M, Rodrigues FP, Medei E, Nascimento JHM. Noninvasive method for electrocardiogram recording in conscious rats: Feasibility for heart rate variability analysis. Anais da Academia Brasileira de Ciências. 2010;**82**:431-437. DOI: 10.1590/s0001-37652010000200019

[8] Chang YT, Wann SR, Wu PL, Hsieh KH, Lin CC, Huang MS, et al. Influence of age on heart rate variability during therapeutic hypothermia in a rat model. Resuscitation. 2011;**82**:1350-1354. DOI: 10.1016/j.resuscitation.2011.04.031

[9] Hashimoto M, Kuwahara M, Tsubone H, Sugano S. Diurnal variation of autonomic nervous activity in the rat - Investigation by power spectral analysis of heart rate variability. Journal of Electrocardiology. 1999;**32**:167-171. DOI: 10.1016/S0022-0736(99)90095-X

[10] Hashimoto M, Harada T, Ishikawa T, Obata M, Shibutani Y. Investigation on diabetic autonomic neuropathy assessed by power spectral analysis of heart rate variability in WBN/Kob rats. Journal of Electrocardiology. 2001;**34**:243-250. DOI: 10.1054/jelc.2001.25130

[11] Mamalyga ML. Circadian changes in cardiac rhythm structure in decompensated chronic heart failure. Bulletin of Experimental Biology and Medicine. 2014;**156**:499-503. DOI: 10.1007/s10517-014-2384-5

[12] Hsieh IT, Yang CC, Chen CY, Lee GS, Kao FJ, Kuo KL, et al. Uninterrupted wireless long-term recording of sleep patterns and autonomic function in freely moving rats. Journal of Medical and Biological Engineering. 2013;**33**:79-86. DOI: 10.5405/jmbe.1039

[13] Uechi M, Asai K, Osaka M, Smith A, Sato N, Wagner TE, et al. Depressed

heart rate variability and arterial baroreflex in conscious transgenic mice with overexpression of cardiac G(s alpha). Circulation Research. 1998;**82**:416-423. DOI: 10.1161/01.RES.82.4.416

[14] Mäenpää M, Penttilä J, Laitio T, Kaisti K, Kuusela T, Hinkka S, et al. The effects of surgical levels of sevoflurane and propofol anaesthesia on heart rate variability. European Journal of Anaesthesiology. 2007;**24**:626-633. DOI: 10.1017/S026502150700004X

[15] Tivesten A, Bollano E, Caidahl K, Kujacic V. The growth hormone secretagogue hexarelin improves cardiac function in rats after experimental myocardial infarction. Endocrinology. 2000;**141**:60-66. DOI: 10.1016/S1096-6374(99)80060-4

[16] Flumignan RLG, Kanashiro RM, Saraiva RM, Portes LA, Antonio EL, Ishigai MMS, et al. Incidence of heart failure in infarcted rats that die spontaneously. Brazilian Journal of Medical and Biological Research. 2006;**39**:1323-1328. DOI: 10.1590/S0100-879X2006001000008

[17] Cohen-Boulakia F, Valensi PE, Boulahdour H, Lestrade R, Dufour-Lamartinie JF, Hort-Legrand C, et al. In vivo sequential study of skeletal muscle capillary permeability in diabetic rats: Effect of anthocyanosides. Metabolism. 2000;**49**:880-885. DOI: 10.1053/meta.2000.6754

[18] Guzzetti S, Marchi A, Bassani T, Citerio G, Porta A. Univariate and bivariate symbolic analyses of cardiovascular variability differentiate general anesthesia procedures. Physiological Measurement. 2015;**36**:715-726. DOI: 10.1088/0967-3334/36/4/715

[19] Fields RD. Vive la différence. Requiring medical researchers to test males and females in every experiment sounds reasonable, but it is a bad idea. Sci Am. 2014;**311**(3):14. PMID: 25211885

[20] Marts SA, Keitt S. Foreward: a historical overview of advocacy for research in sex-based biology. Adv. Molecular and Cellular Biology. 2004;**2004**(34):v-xiii. DOI: 10.1016/S1569-2558(03)34024-X

[21] Wizemann TM, Pardue ML. Exploring the Biological Contributions to Human Health: Does Sex Matter? Washington DC: National Academy Press; 2001. ISBN-10: 0-309-07281-6

[22] Korczeniewska OA, Husain S, Khan J, Eliav E, Soteropoulos P, Benoliel R. Differential gene expression in trigeminal ganglia of male and female rats following chronic constriction of the infraorbital nerve. European Journal of Pain. 2018;**22**:875-888. DOI: 10.1002/ejp.1174

[23] Simoes ALB, Silva GAR, Giorgetto C, do Carmo-Campos ED, Dias FJ, VPS F. Substance P in dorsal root ganglion neurons in young and adult rats, after nociceptive stimulation during the neonatal period. Anat Rec-Adv Int Anat Evol Biol. 2018;**301**:849-861. DOI: 10.1002/ar.23755

[24] Ishii H, Onodera M, Ohara S, Tsutsui KI, Iijima T. Sex differences in risk preference and cFos expression in paraventricular thalamic nucleus of rats during gambling task. Front. Behav. Neurosci. 2018;**12**. article 68. DOI: 10.3389/fnbeh.2018.00068

[25] Hazelhoff MH, Torres AM. Gender differences in mercury-induced hepatotoxicity: Potential mechanisms. Chemosphere. 2018;**202**:330-338. DOI: 10.1016/j.chemosphere.2018.03.106

[26] Svorc P, Petrasova D, SvorcJr P. Chronobiological study of sex differences

in the internal environment in zoletil-anaesthetized rats. Biological Rhythm Research. 2020b;**51**:770-779. DOI: 10.1080/09291016.2018.1564577

[27] Koresh O, Kaplan Z, Zohar J, Matar MA, Geva AB, Cohen H. Distinctive cardiac autonomic dysfunction following stress exposure in both sexes in an animal model of PTSD. Behavioural Brain Research. 2016;**308**:128-142. DOI: 10.1016/j.bbr.2016.04.024

[28] Svorc P, Petrasova D, SvorcJr P. Sex differences in HRV under general anesthesia in rat model. Anesth Pain Res. 2020a;**4**:1-4. DOI: 10.33425/2639-846X.1039

[29] Meyer MR, Haas E, Barton M. Gender differences of cardiovascular disease: new perspectives for estrogen receptor signaling. Hypertension. 2006;**47**:1019-1026. DOI: 10.1161/01.HYP.0000223064.62762.0b

[30] Sgoifo A, De Boer SF, Buwalda B, Korte-Bouws G, Tuma J, Bohus B, et al. Vulnerability to arrhythmias during social stress in rats with different sympathovagal balance. Amer J Physiol-Heart Circ Physiol. 1998;**275**:H460-H466. DOI: 10.1152/ajpheart.1998.275.2.H460

[31] Jiang M, Murias JM, Chrones T, Sims SM, Lui E, Noble EG. American ginseng acutely regulates contractile function of rat heart. Front Pharmacol. 14 Mar 2014;**5**:43. DOI: 10.3389/fphar.2014.00043

[32] Van Buren T, Schiereck P, De Ruiter GJW, Gispen WH, De Wildt DJ. Vagal efferent control of electrical properties of the heart in experimental diabetes. Acta Diabetologica. 1998;**35**:19-25. DOI: 10.1007/s005920050096

[33] Imai K, Ariga H, Chen C, Mantyh C, Pappas TN, Takahashi T. Effects of electroacupuncture on gastric motility and heart rate variability in conscious rats. Autonom Neurosci-Basic Clin. 2008;**138**:91-98. DOI: 10.1016/j.autneu.2007.11.003

[34] Shumilova TE, Shereshkov VI, YanvarevaIN, NozdrachevAD. Peculiarities of myocardial electrogenesis in laboratory rats under conditions of acute nitrite intoxication. Journal of Evolutionary Biochemistry and Physiology. 2010;**46**:179-188. DOI: 10.1134/S0022093010420079

[35] Liu B, Li S, Su Y, Xiong MT, Xu YW. Comparative study of the protective effects of terfenadine and amiodarone on barium chloride/aconitine-induced ventricular arrhythmias in rats: A potential role of terfenadine. Molecular Medicine Reports. 2014;**10**:3217-3226. DOI: 10.3892/mmr.2014.2640

[36] Yamanushi TT, Kabuto H, Hirakawa E, Janjua N, Takayama F, Mankura M. Oral administration of eicosapentaenoic acid or docosahexaenoic acid modifies cardiac function and ameliorates congestive heart failure in male rats. J Nutrit. 2014;**144**:467-474. DOI: 10.3945/jn.133.175125

[37] Abood AM, Elshal MF. VDR stimulation improves outcome of isoprenaline-induced myocardial infarction in rats via down-regulation of cardiac inos gene expression. Biomedical Research. 2015;**26**:755-764. www.biomedres.info

[38] Medei E, Lima-Leopoldo AP, Pereira-Junior PP, Leopoldo AS, Campos DHS, Raimundo JM, et al. Could a high-fat diet rich in unsaturated fatty acids impair the cardiovascular system? The Canadian Journal of Cardiology. 2010;**26**:542-548. DOI: 10.1016/S0828-282X(10)70469-4

[39] Parasuraman S, Raveendran R, Selvaraj RJ. Effects of Cleistanthins A and B on Blood Pressure and electrocardiogram in Wistar Rats. Zeitschrift fur Naturforschung Section c-a Journal of Biosciences. 2011;**2011**(66):581-587. DOI: 10.5560/ZNC.2011.66c0581

[40] Mutiso SK, Rono DK, Bukachi F. Relationship between anthropometric measures and early electrocardiographic changes in obese rats. BMC Research Notes. 2014;**7**(931):3-7. http://www.biomedcentral.com/1756-0500/7/931

[41] Binu P, Priya N, Abhilash S, Vineetha RC, Nair RH. Studies on curative efficacy of monoterpene eugenol on anti- leukemic drug arsenic trioxide induced cardiotoxicity. Biomedicine & Pharmacotherapy. 2017;**91**:559-566. DOI: 10.1016/j.biopha.2017.04.087

[42] Yadav RK, Rawat JK, Gautam S, Singh M, Kumar M, Ansari MN, Roy S, Saeedan AS, Kaithwas G. Antidiabetic activity of mefloquine via GLP-1 receptor modulation against STZ-NA-induced diabetes in albino wistar rats. 3 Biotech. 2018;**8**:240. DOI: 10.1007/s13205-018-1250-y

[43] da Silva VJD, Neto EF, Salgado HC, Junior RF. Chronic converting enzyme inhibition normalizes QT interval in aging rats. Brazilian Journal of Medical and Biological Research. 2002;**35**:1025-1031. DOI: 10.1590/S0100-879X2002000900003

[44] Kralova E, Mokran T, Murin J, Stankovicova T. Electrocardiography in two models of isoproterenol-induced left ventricular remodeling. Physiological Research. 2008;**57**(suppl. 2):583-589. DOI: 10.33549/physiolres.931556

[45] Joukar S, Ghorbani-Shahrbabaki S, Hajali V, Sheibani V, Naghsh N. Susceptibilitytolife-threateningventricular arrhythmias in an animal model of paradoxical sleep deprivation. Sleep Medicine. 2013;**14**:1277-1282. DOI: 10.1016/j.sleep.2013.07.008

[46] Kralova E, Racanska E, Vicenova A, Boselova I, Malik I, Stankovicova T. Pharmacological evaluation of the effects of phenylcarbamic acid derivatives on cardiovascular functions in rats. Acta Pharmaceutica. 2018;**68**:507-515. DOI: 10.2478/acph-2018-0034

[47] Raji-Amirhasani A, Joukar S, Naderi-Boldaji V, Bejeshk MA. Mild exercise along with limb blood-flow restriction modulates the electrocardiogram, angiotensin, and apelin receptors of the heart in aging rats. Iranian Journal of Basic Medical Sciences. 2018;**21**:558-563. DOI: 10.22038/IJBMS.2018.24796.6165

[48] Jain PG, Mahajan UB, Shinde SD, Surana SJ. Cardioprotective role of FA against isoproterenol induced cardiac toxicity. Molecular Biology Reports. 2018;**45**:1357-1365. DOI: 10.1007/s11033-018-4297-2

[49] Sharma S, Khan V, Najmi AK, Alam O, Haque SE. Prophylactic treatment with icariin prevents isoproterenol-induced myocardial oxidative stress via nuclear factor-like 2 activation. Pharmacognosy Magazine. 2018;**14**(supl. S):S227-S236. DOI: 10.4103/pm.pm_469_17

[50] Molcan L, Teplan M, Vesela A, Zeman M. The long-term effects of phase advance shifts of photoperiod on cardiovascular parameters as measured by radiotelemetry in rats. Physiological Measurement. 2013;**34**:1623-1632. DOI: 10.1088/0967-3334/34/12/1623

[51] Molcan L, Vesela A, Zeman M. Repeated phase shifts in the lighting regimen change the blood pressure response to norepinephrine stimulation in rats. Physiological Research. 2014;**63**:567-575. DOI: 10.33549/physiolres.932653

[52] Svorc P, Petrasova D, SvorcJr P. Arterial pH and blood gas values in rats under three types of general anesthesia: A Chronobiological study. Physiological Research. 2018;**67**:721-728. DOI: 10.33549/physiolres.933692

[53] Mortola JP, Seifert EL. Hypoxic depression of circadian rhythms in adult rats. Journal of Applied Physiology. 2000;**88**:365-368. DOI: 10.1152/jappl.2000.88.2.365

[54] Bishop B, Silva G, Krasney J, Nakano H, Roberts A, Farkas G, et al. Ambient temperature modulates hypoxic-induced changes in rat body temperature and activity differentially. Amer J Physiol. 2001;**2001**(280):R1190-R1196. DOI: 10.1152/ajpregu.2001.280.4.R1190

[55] Bosco G, Ionadi A, Panico S, Faralli F, Gagliardi R, Data P, et al. Effects of hypoxia on the circadian patterns in men. High Altitude Medicine & Biology. 2003;**4**:305-318. DOI: 10.1089/152702903769192269

[56] Kaplan JL, Gao E, De Garavilla L, Victain M, Minczak B, Dalsey WC. Adenosine A1 antagonism attenuates atropine-resistant hypoxic bradycardia in rats. Academic Emergency Medicine. 2003;**10**:923-930. DOI: 10.1197/S1069-6563(03)00309-9

[57] Mortola JP. Hypoxia and circadian patterns. Respiratory Physiology & Neurobiology. 2007;**158**:274-279. DOI: 10.1016/j.resp.2007.02.005

[58] Hayashida Y, Hirakawa H, Nakamura T, Maeda M. Chemoreceptors in autonomic responses to hypoxia in conscious rats. Advances in Experimental Medicine and Biology. 1996;**410**:439-442. DOI: 10.1007/978-1-4615-5891-0_67

[59] SvorcJr P, Bačová I, Svorc P, Buzga M. Autonomic nervous system under ketamine/xylazine and pentobarbital anaesthesia in a Wistar rat model: A chronobiological view. Prague Medical Report. 2013;**114**:72-80. DOI: 10.14712/23362936.2014.25

[60] Picker O, Scheeren TWL, Arndt JO. Inhalation anaesthetics increase heart rate by decreasing cardiac vagal activity in dogs. Br J Anaest. 2001;**87**:748-754. DOI: 10.1093/bja/87.5.748

[61] Hoffmann TJ, Simon BJ, Yi Z, Emala CW. Low voltage vagal nerve stimulation reduces bronchoconstriction in guinea pigs through catecholamine release. Neuromodulation. 2012;**15**:527-536. DOI: 10.1111/j.1525-1403.2012.00454.x

[62] Miner JR, Lewis LM, Mosnaim GS, Varon J, Theodoro D, Hoffmann TJ. Feasibility of percutaneous vagus nerve stimulation for the treatment of acute asthma exacerbations. Academic Emergency Medicine. 2012;**19**:421-429. DOI: 10.1111/j.1553-2712.2012.01329.x

[63] Steyn E, Mohamed Z, Husselman C. Non-invasive vagus nerve stimulation for the treatment of acute asthma exacerbationsresults from an initial case series. International Journal of Emergency Medicine. 2013;**6**:7-9. DOI: 10.1186/1865-1380-6-7

[64] Yuan H, Silberstein SD. Vagus nerve and vagus nerve stimulation, a comprehensive review: Part III. Headache. 2016;**56**:479-490

[65] Shi S, Liu T, Wang D, Zhang Y, Liang L, Yang B, et al. Activation of N-methyl-D-aspartate receptors reduces heart rate variability and facilitates atrial fibrillation in rats. Europace. 2017;**19**:1237-1243. DOI: 10.1093/europace/euw086

[66] Damasceno DD, Lima MP, Motta DF, Ferreira AJ, Quintão-Junior JF, Drummond LR, et al. Cardiovascular

and eletrocardiographic parameters after tonin administration in Wistar rats. Regulatory Peptides. 2013;**181**:30-36. DOI: 10.1016/j.regpep.2012.12.009

[67] Yadav RK, Rawat JK, Gautam S, Singh M, Kumar M, Ansari MN, et al. Antidiabetic activity of mefloquine via GLP-1 receptor modulation against STZ-NAinduced diabetes in albino wistar rats. Journal of Biotechnology. 2018;**8**:240-250. DOI: 10.1007/s13205-018-1250-y

[68] Svorc P, Petrasova D, SvorcJr P. Heart rate variability and heart rate under general anesthesia in rats of both sexes. Trends in Medicine. 2020c;**21**:1-3. DOI: 10.15761/TiM.1000257

[69] Pagani M, Lombardi F, Guzzetti S, Rimoldi O, Furlan R, Pizzinelli P, et al. Power spectral analysis of heart rate and arterial pressure variabilities as a marker of sympatho-vagal interactions in man and conscious dog. Circulation Research. 1986;**59**:178-193. DOI: 10.1161/01.RES.59.2.178

[70] Pagani M, Lombardi F, Guzzetti S, Sandrone G, Rimoldi O, Malfatto G, et al. Power spectral density of heart rate variability as an index of symptho-vagal interactions in normal and hypertensive subjects. Journal of Hypertension. 1984;**2**:383-385

[71] Malliani A, Pagani M, Lombardi F, Cerutti S. Cardiovascular neural regulation explored in the frequency domain. Circulation. 1991;**84**:482-492. DOI: 10.1161/01.cir.84.2.482

[72] Billman GE. The LF/HF ratio does not accurately measure cardiac sympatho-vagal balance. Frontiers in Physiology. 2013;**4**:26. DOI: 10.3389/fphys.2013.00026